BLOODSTAIN PATTERN EVIDENCE

BLOODSTAIN PATTERN EVIDENCE: OBJECTIVE APPROACHES AND CASE APPLICATIONS

ANITA Y. WONDER

AMSTERDAM • BOSTON • HEIDELBERG • LONDON
NEW YORK • OXFORD • PARIS • SAN DIEGO
SAN FRANCISCO • SINGAPORE • SYDNEY • TOKYO

Academic Press is an imprint of Elsevier

Acquisitions Editor:	Jennifer Soucy
Assistant Editor:	Kelly Weaver
Project Manager:	Christie Jozwiak
Marketing Manager:	Diane Jones

Elsevier Academic Press
30 Corporate Drive, Suite 400, Burlington, MA 01803, USA
525 B Street, Suite 1900, San Diego, California 92101-4495, USA
84 Theobald's Road, London WC1X 8RR, UK

This book is printed on acid-free paper. ∞

Library of Congress Cataloging-in-Publication Data
APPLICATION SUBMITTED

British Library Cataloguing in Publication Data
A catalogue record for this book is available from the British Library

ISBN 978-0-12-370482-5

For all information on all Elsevier Academic Press publications visit our Web site at www.books. elsevier.com

Printed and bound by CPI Group (UK) Ltd, Croydon, CR0 4YY

Transferred to Digital Print 2011

Dedication

For Brian Parker, D.Crim., J.D.
Professor Emeritus California State University,
Sacramento, for building up my confidence
and calming down my zeal, and most importantly,
for knowing which was necessary.

TABLE OF CONTENTS

FOREWORD ix

ACKNOWLEDGMENTS xi

SECTION I: INTRODUCTION 2

    Chapter 1:   Introduction 3

    Chapter 2:   The American Founding Father of Bloodstain Pattern
                       Evidence—The Kirk Years 17

    Chapter 3:   Understanding Blood Behavior 23

    Chapter 4:   Trigonometry in Bloodstain Pattern Evidence,
                       Math Use in Question 33

SECTION II: CASE APPLICATIONS 48

    Chapter 5:   Timing is Everything 49

    Chapter 6:   The Alexander Lindsay Second Inquiry 63

    Chapter 7:   Who Was the Shooter? 79

    Chapter 8:   Traffic Salutations in America 87

    Chapter 9:   The Body on the Porch 95

    Chapter 10: Lil' Ol' Guy Who Woke Up Dead 101

    Chapter 11: Self-Defense Staging Homicide by Gunshot 111

    Chapter 12: Crime Scene Artwork, Staged Assault 123

    Chapter 13: Curiosity Caught the Murderer 135

    Chapter 14: Informant Execution 141

    Chapter 15: Lack of the Crime Lab Involvement 147

    Chapter 16: On Duty Officer Involved Shooting 155

    Chapter 17: Three Down and Still Missed Intended Victim 161

    Chapter 18: Perfect Stories 167

    Chapter 19: Family Elimination 173

Chapter 20:   A Tragedy of Errors Homicide   183

Chapter 21:   The Accomplice Wouldn't Plea   189

Chapter 22:   Magic Bullet in Alleged Drive-By Shooting   197

Chapter 23:   Perry Mason Is a Myth   203

Chapter 24:   Hidden Face, Blunt Force Assault   209

Chapter 25:   Body in the Bathtub   213

SECTION III: BLOODSTAIN PATTERN EVIDENCE INTERACTIONS WITH OTHER FORENSIC DISCIPLINES   218

Chapter 26:   Bloodstain Pattern Evidence and Law Enforcement   219

Chapter 27:   The Forensic Crime Lab and Bloodstain Pattern Evidence   237

Chapter 28:   DNA and Bloodstain Pattern Evidence   251

Chapter 29:   Pathology and Bloodstain Pattern Evidence: The Predominant Good,
               Occasional Bad, and Rare Ugly   257

Chapter 30:   Bloodstain Pattern Evidence and the Law   265

SECTION IV: EXPERIENCE WITH TRAINING IN BLOODSTAIN PATTERN EVIDENCE   276

Chapter 31:   Preparations for Bloodstain Pattern Workshops   277

Chapter 32:   Spatter Group Exercises   287

Chapter 33:   Exercises with Groups Other Than Spatters (Nonspatter Groups)   303

Chapter 34:   Special Projects, Practical Exams, and Mock Crime Scenes   315

SECTION V: THE FUTURE OF BLOODSTAIN PATTERN EVIDENCE   326

Chapter 35:   Research in Bloodstain Pattern Evidence   327

Chapter 36:   Summation   345

APPENDIXES   352

Appendix A: Flow Diagram   353

Appendix B: An Objective Approach to Spatter Classification Based on SAADD   355

Appendix C: Spatter Classification Table   361

Appendix D: Twenty-five Bloodspatters to Practice Measuring   363

Appendix E: Tips on Sequencing Pattern Categories   365

Appendix F: Some Random Values for Future Research   367

Appendix G: Glossary   369

Appendix H: List of Abbreviations   375

INDEX   377

FOREWORD

FOREWORD – PRINCIPLES AND PRACTICE

This follow-up book to *Blood Dynamics* resembles the first in that it challenges some of the holy grails of blood pattern analysis from a sound scientific basis. It provides a deeper understanding of some of the principles which need to be adhered to if investigators are not to be mislead, and we are to be assured of safe and sustainable courtroom verdicts based on blood pattern evidence. This reflects Anita Wonder's unusually deep understanding of blood and all its variations, which she developed originally as a microbiologist and health care haematologist and then as a blood pattern expert and trainer of considerable reputation in the forensic world.

For the first time we have a clear exposition about why there should be such lack of clarity about the terms used to describe blood patterns and which have caused so much confusion and unnecessary debate and disagreement in the past. Far from being prescriptive, Anita is merely concerned that we should understand the principles underlying the terms we use so that those of us in the field can communicate effectively with one another and with the end users of our services.

As always, she encourages and helps her readers to rely upon analyzing individual blood stains and groups of stains and to use these to suggest scenarios, and only then to consider how they might impact on the specific context of the case. The common alternative—to develop a scenario and then explore to what extent the findings might fit it, tends to provide a sort of 'fits where it touches' type of evidence which is often deeply flawed. Also amply demonstrated is the importance of experimentation—not in a simplistic, pattern matching kind of way, but to promote a deeper and unbiased understanding of how patterns have been formed and the sorts of insights this can provide for particular case circumstances.

All in all, this book is a very valuable addition to the pantheon of blood pattern analysis literature, contributing substantially to our necessary understanding of the technical background which should underpin our efforts. It not only informs but it stimulates and stretches us, causing us to question previously held beliefs and practices. Surely this is precisely what true scientists should do and, in this, Anita has been supremely successful.

ANGELA GALLOP
January 2007

ACKNOWLEDGMENTS

The usual comments that a book is never the work of one person is very much true of this one. Further, my experiences and opinions are based upon the feedback, discussions, and contributions of many well-qualified people with whom I have been fortunate in coming into contact with over the years. First and foremost is G. Michele Yezzo, who has been my sister in spirit, business partner, colecturer, advisor, critic, and friend for 20 years. Her input on *Blood Dynamics* and *Bloodstain Pattern Evidence* has been essential.

The urging and moral support from Sgt. Warren Day, Sgt. Dennis Dolezal, and Dr. Angela Gallop made this a reality. Each in their own way kept me focused and aware of the needs of the people who do the work. Sgt. Dean Reichenberg has been there to support me in my many experiments and his professionalism continues to impress all who work with him. And I will always have fond memories of the Bloody Workshop crews. In *Blood Dynamics* I forgot to mention James Conley of the Anaheim Police Department. Jim was a delightful critic and assistant in the workshops and provided valuable feedback from the detective viewpoint with science education (he majored in physics in college).

Not only have I been blessed with many excellent practical advisors, but I've been uncannily lucky at coming into contact with supreme academic advisors, including Richard Saferstein, John I. Thornton, Brian Parker, Talib ul Haq, George Roche, Angela Gallop, and law school professor Edward Imwinkelried. Medical doctors have been great at keeping me on track regarding physiology and anatomy and realistic regarding injuries and autopsy findings. My first experience with a truly great forensic pathologist was with Dr. Pierce Rooney in Sacramento, CA, who was also one of my first clinical pathologists. Dr. Gwyn Hall was doing a forensic residency when I met her, and she was as enthusiastic in teaching me as in learning herself. After such a good beginning I met many others who were helpful or not, as was their natures, but in later years my best experience has been with the delightful Canadian Dr. Chitra Rao.

Last but certainly not least, the lawyers. My experience with attorneys has been predominantly good. Of course I've encountered rare individuals who want to tell me what I should say as an expert. Unfortunately they are able to do that with the experts they use, and therefore continue. All the attorneys who have hired me for the full run of a case have been more interested in learning what bloodstain pattern evidence can tell them regarding the events involved. The attorney who was the most open to input and understood the underlying science is now His Honor Judge John Nicholson in Australia. John is still willing to advise and correct a legally inexperienced scientist. Of course the best law instructor anyone could have is Ed Imwinkelried of UC Davis Law School. I've learned that the more you know the more precise the answer to any question.

Addendum, gratitude is due to my editor Jennifer Soucy and assistant editor Kelly Weaver. When I was given this assignment, my thoughts were "piece of cake" since *Blood Dynamics* was completed in about three months. The difference between a 100 page book and a 500 page book is legion. I couldn't have completed it without the guidance and communications from Jennifer and Kelly.

Bloodstain pattern evidence is a science, but science applications presently are underutilized in this discipline. Hopefully this book and those being written, and the great minds who will read them, will enhance the process for expanding knowledge in the future.

ANITA Y. WONDER
January 2007

SECTION I

INTRODUCTION

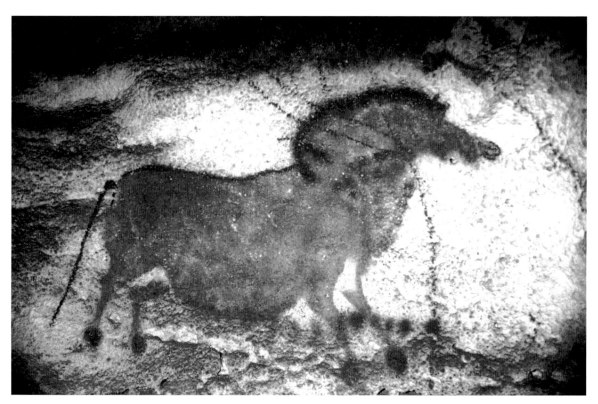

Figure 1-1   "How to Kill a Horse." A 17,000-year-old cave painting at Lascaux, France. Reprinted by permission of the French Ministry of Culture. The red streaks have been identified by French scientists as vegetation (possibly a tree). From the perspective of bloodstain pattern analysts it resembles a training session, or observations where arterial damage occurred to a wild horse.

CHAPTER 1

INTRODUCTION

OBJECTIVES OF SECOND VOLUME

The publication of *Blood Dynamics*[1] broke with tradition in bloodstain pattern evidence (BPE) in that the emphasis shifted from following early pattern match exercises toward a more technical base gleaned from science disciplines not directly related to forensics. This is consistent with the application of most of what we now include within the field of criminalistics. Pure science research provides the foundation upon which development and incorporation of procedures, techniques, and interpretations follows for forensic science adapted protocols. Because BPE is relatively new within recognized science disciplines, both subjective and objective approaches presently exist in casework, training, and adjudications. This volume is written as an expanded follow-up to *Blood Dynamics*.

OBJECTIVES FOR BLOODSTAIN PATTERN EVIDENCE, OBJECTIVE APPROACHES AND CASE APPLICATIONS:

1. Expand upon the science information introduced in *Blood Dynamics*.
2. Illustrate an objective approach to bloodstain pattern analysis (BPA) in casework.
3. Offer views from different forensic perspectives and applications of the evidence.
4. Discuss training techniques and experiences from over 30 workshops.
5. Offer suggestions toward research in this discipline.

Bloodstain Pattern Evidence is now frequently included in science presentations, training programs, and new Forensic Science publications. This indicates that readers are interested in gaining more information regarding the investigative potential of what is perhaps history's oldest form of physical evidence. In *Blood Dynamics*,[2] it was suggested that Paleolithic humans, 15 millennia ago, recognized bloodstain patterns as associated with mortality. Respiratory (exhalation, expiration) type blood drop distribution may be interpreted from the spots added around the head of *Wounded Bison*, a cave painting at Alta Mira,

---

[1]Wonder, Anita Y. (2001). *Blood Dynamics*. AP Forensics, London.
[2]*Ibid.*, 37.

Spain. Another interesting pattern is seen in the 17,000-year-old cave painting at Lascaux, France (see Figure 1-1), which resembles the dynamics and resultant bloodstain patterns from arterial damage (arterial spurts, gush). Blood flows from apparent injuries (Figures 1-2 and 1-3) were carved in the walls at the ancient city of Ninehvey (present-day Kuyunjik, Iraq) dating back 2650 to 2645 years.[3] Swing cast offs (cast offs) appear on a modern facsimile of an early thirteenth century Hawaiian petroglyph (Figure 1-4). Despite this ancient and global history, acceptance, classification, application, and training are still open for review and revision with BPE.

Recently forensic science emphasis has focused on standardization. With specialties such as DNA, fingerprints, fibers, drug assays, firearms, and tool

Figure 1-2

Dying Lion *of Nineveh (Iraq 650 BC).* © Trustees of the British Museum.

Figure 1-3

Dying Lioness *of Nineveh (Iraq 645 BC).* © Trustees of the British Museum.

marks, standardization is essential and straightforward to plan. Complete standardization may be more difficult to define with bloodstain pattern evidence, because of the scope of the discipline. Areas that may be involved overlap with other defined departments and may include (but are not limited to) spatter events, bloody transfers, coagulation information, flows with drying, clothing examination, medical wound and injury reports, firearms dynamics, photography, legal preparation, public opinions,

[3]*Dying Lioness* and *Dying Lion.* © Copyright the Trustees of the British Museum.

and various technical subjects such as physics, mathematics, colloidal chemistry, and engineering principles.

It may also be too early to attempt to standardize some aspects of BPE. The most recent scientific discoveries regarding blood behavior have not been, at this point in time, incorporated into understanding the discipline. Furthermore, presently the ultimate decision of whether the evidence will be used in investigations and adjudication lies not with science professions, but rather with law enforcement officers and attorneys. Unfortunately subjective applications may sometimes inadvertently be favored over objective ones when the evidence is presented. It would be a mistake for the future of a science to standardize principles at a time when subjectivity may prevail.

Figure 1-4

Facsimile of stone petroglyph found in areas of the Hawaiian Islands.

In *Blood Dynamics*,[4] the objective was to move toward a more scientific explanation of blood behavior using references from a variety of science fields. No effort was directed at censoring anyone engaged in the use of the evidence. The emphasis has always been to point out updated science principles, and to encourage everyone to broaden their perspective of the range of potential which bloodstain pattern evidence offers. This book will continue the technical approach begun in *Blood Dynamics*, and include practical applications in actual casework. The cases presented show one way the information could be developed. There is no claim that this is the only way to apply the information. Learning is continuous with each exposure to cases, students, and other experts. Bloodstain pattern evidence is the ultimate in forensic continuing education.

Because this work is designed to bring various viewpoints together, rather than approach the material from a single perspective, different writing styles are incorporated here. The objective is to provide useful information for a widely divergent audience using semantics from a range of experience and academic backgrounds. Hopefully this format can also act to bring together the many individuals who are involved with and needed in the field of bloodstain pattern analysis (BPA). The reader should check the appendixes for guidelines in approaching case material and understanding abbreviations for terminology used in this book.

---

[4]Wonder, *Blood Dynamics*, 5.

TERMINOLOGY GENERAL CONSIDERATIONS

Terminology is a source of divergent viewpoints within organizations with members who practice bloodstain pattern analysis. Attempts to standardize terms have been ongoing since the formation of the International Association of Bloodstain Pattern Analysts (IABPA) in 1983, an organization with approximately 80 percent law enforcement membership. The benefits of a set terminology list cannot be denied. However, there is a problem with one list for BPE. The involvement of professionals with different viewpoints may require more than one definition for some terms. Since medical science is part of the full analysis, perhaps a medical analogy will help clarify the need for alternative semantics.

If blood thinner is mentioned, most people will recognize the phrase. It is found in newspapers, magazines, and instructions given to patients with problems in blood circulation. In fact, there is no such thing as a blood thinner. The term is a colloquial expression for the benefit of people who need to take medication to prevent their blood from clotting within blood vessels, i.e., a "thinner" to prevent "thickening." This term is not used in clinical laboratory science. Instead, analysis is done on plasma to test for circulating anticoagulants. Physicians who treat bleeding disorders, however, do not use the term anticoagulant, as it is deemed too narrow for their purposes. Hematology specialists use the phrase hemostasis inhibitors.[5]

An analogy to the situation with bloodstain pattern terminology would be if a group of patients, who greatly outnumber technologists and doctors, voted to only use the phrase blood thinners when discussing prevention of coagulation. The refusals to comply by physicians and laboratory scientists would be analogous to that which sometimes occurs with attempts to establish a required terminology list derived from the law enforcement history of bloodstain pattern analysis.

The discipline is a focus of law enforcement. There is no denying the importance of good police work involving early recognition of BPE at a crime scene, and applying careful and competent bloodstain pattern analysis. Law enforcement agencies, and agencies dealing predominantly with law enforcement, however, have their own specialized language. Cop speak[6] is essential in intradepartmental communication. Although ingrained and necessary, these terms should not be required as the only acceptable terms for all involved with BPE. Bloodstain pattern evidence, unlike DNA, toxicology, and other technical specialties, will always be split between law enforcement and the forensic science laboratory. For the maximum benefits to each, common ground is essential with mutual respect for communication between different levels of experience and academics with consideration for scientific labeling.

---

[5]Rosenberg, Robert D., MD, PhD. (1987). Regulation of the hemostatic mechanism. In *The Molecular Basis of Blood Diseases*. W.B. Saunders Co., Philadelphia, 534 f.
[6]Bugliosi, Vincent. (1996). *Outrage*. W.W. Norton & Co., New York.

TERMINOLOGY DISCUSSION AND REVIEW FROM *BLOOD DYNAMICS*

The cases presented in Section II include terms used at the time of analysis and some developed for clarity later. Many terms were from a traditional earlier lexicon, and a few were derived for better understanding during class presentations. The first use of a pattern label favored by the author is presented in bold type and a short definition is provided. Other applicable terms known to be in use follow in parentheses. It is the intent of this work that analysts accept the existence of variation in semantics. Because understanding how patterns are formed, and why they appear as they do, is so essential to identification, the focus of this work is on how terminology defines the evidence rather than as a memorized list. To further this approach the glossary and margin definitions may vary.

Figure 1-5

Schematic example of the dynamics in drip cast offs. Manikin® Courtesy of NexGen Ergonomics Inc. (www.nexgenergo.com, www.humancad.com).

Hopefully, in the future, various applicable committees will focus on what is right with regard to science principles rather than who is right as the originator of a term.

Blood Dynamics introduced a simplified approach using a flow diagram of events and the types of patterns that result for each as recognizable divisions of BPE. An updated version is presented in Appendix A. The main purpose of any flow diagram is to show relationships between actions or events. For example, cast offs are uniformly accepted as an event where a moving carrier—material, object, or weapon—is sloughing or casting off blood drops during travel. Drip cast offs are the same action with motion slowed (Figure 1-5). Swing cast offs result from the motion increased in velocity (Figure 1-6), and cessation cast offs occur when the carrier abruptly stops, causing lightly adhering blood to separate and continue flight as drops traveling in the direction the carrier was moving immediately prior to movement cessation (Figure 1-7). On a flow diagram, cast offs is the pattern category with drip, swing, and cessation as subcategories identifying at which relative stage in time and space the blood drops separated from the carrier.

Refer to Appendix A for the full flow diagram.

The three stages of cast offs often are associated within a sequence of acts during a crime. For example, drip cast offs (low velocity impact spatter (LVIS), blood trails, passive, gravitational drops) may exist at the beginning of a swing with a bloodied weapon followed by swing cast offs as a blow is delivered to a victim. These patterns may be associated with the victim throwing up their arm in self defense to block the blow. A defensive gesture may distribute swing cast offs as well as cessation cast offs. The weapon stopping from contact with a raised arm can distribute cessation cast offs. After an assault, drip cast offs (LVIS, blood trails, passive, gravitational stains) may lead from the attack to where the body collapsed or was abandoned, or the weapon was discarded.

Impact spatters: Blood spots resulting from some form of impact dynamics.

Impact spatters identify a single event in time and space, whereas cast offs and arterial damage stains describe the events over a range of time and space, including how an assailant may have used the weapon and a victim moved after injury. One way to remember the three main representatives of the spatter group in terms of investigative leads information is with the generalization that arterial damage shows the actions and movement of the victim; cast offs show the actions of the assailant, and impact spatters show at what origin in space the two came together. Exceptions occur, like the victim dripping drip cast offs (not associated with the assailant), and the assailant having arterial damage patterns on their person (not directly associated with the victim's movement), but the consideration for association is a starting place to identifying the variety of overlapping patterns which are often present at a crime scene.

Figure 1-6

Schematic example of the dynamics in swing cast offs. Manikin® Courtesy of NexGen Ergonomics Inc. (www.nexgenergo.com, www.humancad.com).

VELOCITY IMPACT SPATTER TERMS

A category of patterns historically has been described on the basis of velocity. This description began with Dr. Paul Kirk (1902–1970),[7] Professor of Criminology at the University of California at Berkeley, and was used in much of his pioneering work, including the *amicus curiae* defense brief following the conviction of Sam Sheppard.[8] Dr. Kirk's concept was that drops of blood will leave different shaped stains depending upon how fast the drops were traveling

---

[7]Wonder, *Blood Dynamics*, 3.
[8]Kirk, P.L. Affidavit Regarding State of Ohio vs Samuel H. Sheppard. Court of Common Pleas, Criminal Branch, No. 64571, 26 April 1955.

at the moment they came into contact with a recording surface (target).[9] A drop of any moving liquid will continue traveling after contact with a surface in order to dissipate all forward momentum. The time it takes to dissipate momentum will depend upon the size of the drop (or the mass) and the velocity that the blood drop was traveling at contact with the target. The faster a drop travels and the greater the size of the drop, the more elongated the bloodstain left. This may be applied visually to blood streaks (and/or exclamation marks) found at the scenes of hand gun assault. The smaller the drop, and the less volume and/or less velocity, will result in air friction stopping a blood drop forward travel sooner. This latter concept may be seen in workshops with the spring trap device where small round stains are seen among medium-sized elliptical stains. The spring trap is considered a reproduction of bludgeoning.

Figure 1-7

Schematic example of the dynamics in cessation cast offs. Manikin®️ Courtesy of NexGen Ergonomics Inc. (www.nexgenergo.com, www.humancad.com).

No statements were found in Dr. Kirk's work that associated a specific size blood drop as only occurring from a specific velocity drop. Crime events were described as arrays of drops distributed with whole pattern, group characteristics, not on the basis of a single blood drop. Dr. Kirk pointed out in his Cohn seminars before the California Criminal Trial Lawyers' meetings in San Francisco[10] that it is "important to determine the character of the propelling force for a group of blood spots." In fact, Dr. Kirk further clarified his meaning in a lecture before the California Bar Association Annual Criminal Trial Lawyers' Meeting in San Francisco in 1968.[11] In that program he described the difference between cast off, impact, and arterial distributions as distinctive groupings of blood drops that leave identifiable patterns. Velocity terminology was not included in the notes of that lecture. Copies of his correspondences to, as well as from, his colleagues are kept in files available at the Bancroft Library at the University of California at Berkeley, verifying that others understood his meanings.

Unfortunately, as former students of Dr. Kirk claim, his lectures were not revealing regarding his knowledge or approach in casework with bloodstain

---

[9]Kirk, Paul L. (1967). *Blood—A Neglected Criminalistics Research Area*, Law Enforcement Science and Technology, S.A. Yefsky, Ed. Academic Press, London.
[10]The Paul Kirk papers at the Bancroft Library, UC Berkeley, Box 1 72/55c, 5.
[11]Kirk, Paul L. (1968). *Blood Spot Analysis*. Notes of lecture for the Fourth Annual Criminal Law Seminar, San Francisco. Paul Kirk Papers, UC Bancroft Library, Berkeley, California.

patterns. Our understanding of his logic now must be acquired by review of his correspondence, lectures before trial lawyers, and his cases. In those it is clear that although he regarded the evidence as position and pattern format, he did not advocate memorization of pattern appearance from his simplified reconstruction experiments. His study involved understanding how different dynamics distributed groups of drops that were recorded on surfaces in recognizable arrangements at crime scenes. He realized the importance of the dynamic acts themselves, with regard to directionalities indicated by the groups of drops distributed. More will be discussed regarding this in Section II.

MIST SPATTER AND RESOLUTION

The definition of mist has become ambiguous. Many investigators use the term as synonymous with gunshot distributed impact spatter (GDIS, blow back, HVIS). One expert stated that they found mist 6 feet from the origin of a gunshot homicide; others claim mist does not travel more than 6 inches, or another claims 3 feet. The stains noted for each were visible to the unaided eye, as no microscopic examination was mentioned in the cases. Gunshot distributed impact spatter

Figure 1-8

Example of GDIS (gunshot distributed impact spatters).

Exit wound spatter: Blood drop array distributed from a gunshot exit wound.

(GDIS, HVIS, blow back, forward spatter) fans outward in a rough cone shape from the origin (for either entrance or exit) of the shot. Close to the entrance wound, a pattern resembling spray painting may be seen (Figure 1-8). As the blood drop array is projected away from the origin, individual drops separate and are deposited separate from the group. Single spatters less than 0.1 mm in diameter are not visible alone. What looked like a spray-painted mark when the bloodspatters were close together becomes invisible to the unaided eye when drops are spaced apart. Claims that mist does not travel far from the origin cannot be substantiated unless a thorough microscopic examination was made of the target.

An example of mist ability to travel distances occurred during a firearms experiment. A gunshot, from a .40 Smith and Wesson Model 4013TSW loaded with Remington copper jacketed hollow points, was discharged into a thin plastic sandwich bag containing human, high-ratio (86% hematocrit) blood, which resulted in mist-sized spatter (less than 0.1 mm) projected almost 13 feet. A notebook left on a table 10 feet beyond the bullet exit, **exit wound spatter** (forward spatter) position, and 8 feet to the side of the bullet exit line (12.8 feet on a diagonal from the shot) appeared to have only fine (0.6–0.9 millimeter) and small stains (1–3 mm) recorded (Figure 1-9). When examined with a magnifying loupe, mist-sized stains less than 0.1 mm diameter were found (enlargement seen in Figure 1-10).

In defining bloodspatter size ranges, consideration should be given to the subject of resolution. The traditional definition for the term resolution is the ability to perceive two very small dots close together as separate dots. The limitations of an individual's ability to see dots is influenced by many things including light, texture and color of the surface on which the dots are registered, distance the eye is from the dots, as well as the ability of the eye to focus and detect images. In a very general way, it can be taken that 1.0 mm is visible to the normal eye under favorable conditions. Objects that are less than 0.1 mm would definitely be below visibility, and 0.1 mm to 1.0 mm borderline and variable visibility for most observers under favorable conditions. This can provide rough limits to our terms regarding bloodstain size.

Mist is variously defined as less than 1 mm, less than 0.5 mm, and less than 0.1 mm. It is advisable to clarify the range because stains of 0.6 to 0.9 mm can be found at beatings, knifings, respiratory projected patterns, and satellites from rapid or explosive blood into blood. Most important, the individual stains are visible to the unaided eye. Defining mist as less than 0.1 mm clarifies the need for a spray-paint effect, or an increased spatter density, so that the pattern becomes visible when individual spatters would not be.

Spatters greater than 0.1 mm but less than 1 mm may or may not be visible depending upon conditions, and are classified as fine-sized spatter. Stains greater than 1 mm but less than 3 mm are uniformly visible by anyone capable of analyzing bloodstain patterns, but they are perceived as small, therefore classified as small spatters. Medium becomes 3 mm to 6 mm and large spatters are anything over 6 mm. These are suggestions based upon resolution but subject to change when and if review is accepted by those using the terms and is published from standardization committees. The limits usually apply to the greatest measurement of an elliptical stain, but exclamation mark stains are never measured, and (author's guidelines) bloodstains with calculated incident angles of less than 10 degrees should not have an angle recorded for them (i.e., record as less than 10 degrees).

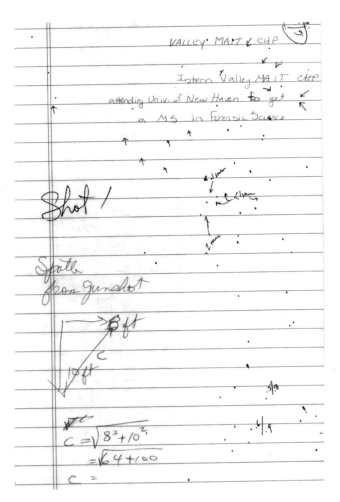

Figure 1-9

Notebook page exposed, horizontally, at gunshot experiment, 10 feet behind and 8 feet to the left of the bullet exit through a plastic sandwich bag containing 5 cc of packed human RBC.

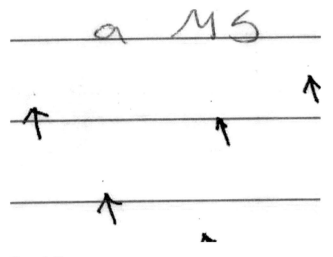

Figure 1-10

Enlargement to show mist <0.1 mm at the tips of arrows.

OTHER TERMS USED IN THE FLOW DIAGRAM

Most of the terms listed on the flow diagram are familiar to those who have had some kind of lecture or discussion with others informed in bloodstain pattern analysis. A few noteworthy differences should be explained. One is the use of the term *expiration* to mean bloodspatters distributed by respiratory functions. Those who use *expiration* often do so to correct the earlier misuse of the word *inspiration* to mean breathing, wheezing, sneezing, and coughing. The latter term was common in forensic evidence reports prior to 1984. It was pointed out to various individuals and instructors that *in* is the taking in and *ex* is the giving out; in-spiration was thus changed to ex-piration.

Taber's Cyclopedic Medical Dictionary[12] uses expiration as a base for definitions of cough and sneeze, but the semantics originated considerably earlier than 1977. The *American Heritage Dictionary* lists the term *expire* as to come to an end, to die. Thus expiration in modern usage identifies death where the victim is incapable of respiration. The old term expire was to breath out, but exhale is used in current lexicon. An alternative form could be respiration or respiratory distribution. Exhalation was used in the cases that follow. Medical students, attending physicians, and Emergency Room personnel all claim to identify the act with common words such as breathing, wheezing, sneezing, and coughing.

The term *void* was discussed in *Blood Dynamics.*[13] *Blockage* and *absence* are favored here because the two types of patterns characterized by lacking bloodstains have different interpretations and investigative lead functions. Void is associated with older, pre-1940s' lexicon.

Blockage patterns are where an obstruction prevents blood flows or drops from reaching a surface, thus may outline a specific obstruction (Figure 1-11). By contrast, absence patterns result from the angle and manner of blood drop distribution, i.e.; cone or megaphone-shaped distribution from firearms, with no obstruction blocking their flight. Notice the effect of the gunshot entrance wound, seen in Figure 1-12, or as seen in the water spray on the asphalt drive in Figure 1-13 (see page 14). Figure 1-14 presents a challenge in classification. Would blockage or absence fit the perpendicular space facing the viewer between the two spattered surfaces?

Blockage template pattern: A blockage transfer where the edges are well defined.

---

[12]*Taber's Cyclopedic Medical Dictionary, 13e.* (1977). Clayton L. Thomas, Ed. F.A. Davis, Philadelphia.
[13]Wonder, *Blood Dynamics*, 84.

BLOODSTAIN PATTERN EVIDENCE IS NOT A PATTERN MATCH EVIDENCE

Even some experienced forensic scientists make the mistake of calling bloodstain patterns a pattern match evidence. Simple direct transfers and some **blockage template patterns** may be classed as pattern match. Spatters from impacts, cast offs, arterial damage, respiratory distribution, and/or blood into blood should not be treated as pattern match evidence. The cases that follow in Section II emphasize that patterns are identified by appearance and grouping criteria, listed in *Blood Dynamics*[14] as:

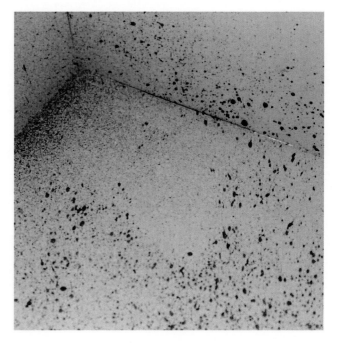

- ▪ **S**hape of the whole pattern
- ▪ **A**lignment of the individual stains with regard to the whole pattern
- ▪ **A**lignment of individual spatters with regard to other spatters
- ▪ **D**istribution of size ranges
- ▪ **D**istribution of the number of spatters

Seasoned investigators learn to apply these without specific thought, and perhaps that has led to the assumption that identification is by pattern match. Experienced examiners can look at an arrangement of spatters and usually identify the probable event that distributed them. Unfortunately new participants of workshops do not immediately have this ability and may feel that they too should be able to just look at patterns and know the interpretation. They try to memorize workshop patterns and fail. This leads to a majority admitting they can't do it, thus losing the benefit of a 40-hour workshop. A few in-

Figure 1-11

Blockage transfer.
Note the shape of a pair of glasses among the spatters from a spring trap device. Two events with removal of glasses in between them.

Figure 1-12

Spatters would be recorded had not an obstruction prevented them from reaching the shooter's forearm. The obstruction could also have been a torn sleeve or a second person grabbing the wrist of the shooter.

---

[14]Wonder, *Blood Dynamics*, 33–34.

Figure 1-13

*Absence patterns can
be seen on each side of
the water on the drive.
The shape of distribu-
tion determines the shape
recorded, not the presence
of an obstruction.*

Figure 1-14

*How would you label
the various surfaces
seen in this picture, and
more important, what
information would that
classification provide?
This is discussed further
in Section IV.*

dividuals feel they learned what they knew
all along based on having seen hundreds
of bloody crime scenes. Neither is a desired
training result. Students seen pantomiming
actions in relation to patterns are appreciating
the three-dimensional and dynamic nature of
the science.

Because of the prevalence in thinking of BPE
as a pattern match forensic science, continued
vigilance is necessary in reading the case examples. Except for transfer evidence, all patterns
are identified by using criteria, while acquiring experience. Continued application of this
technique helps develop instant recognition
of pattern types at crime scenes. A scheme for
identifying spatter patterns is presented in the
appendixes. Refer to these aids while reading Section II. A technique which may
be used is to ask why an initially identified pattern is not one of the other four
major categories, i.e., a recognized impact spatter pattern is analyzed to answer
why it could not be a cast off, arterial, respiratory, or blood into blood pattern.

Before presenting case examples, a glimpse of the man who was responsible for bloodstain pattern evidence in the United States, Paul Leland Kirk,

is presented by the
person most quali-
fied to describe him,
John I. Thornton.

Figure 2-1 Paul Leland Kirk (1902–1970). Father of bloodstain pattern evidence in the United States of America.

CHAPTER 2

THE AMERICAN FOUNDING FATHER OF BLOODSTAIN PATTERN EVIDENCE—THE KIRK YEARS[1]

John I. Thornton

Dr. Paul Leland Kirk will almost certainly be remembered as one of the most notable criminalists of the twentieth century. Initially trained as a chemist and then as a biochemist, he then trained himself to be an innovative microchemist, and later still, applied his formidable intellectual gifts to the discipline we now call Criminalistics.

Paul Kirk was born in Colorado Springs in 1902, the youngest of three children. His early childhood, however, was spent in Pennsylvania.

I imagine it would be a surprise to many of the people who knew him, but—he was a preppy. He attended a prep school, the Randolph Macon Academy at Pt. Royal, Virginia.

His baccalaureate degree was in chemistry from Ohio State University, and his master's was in chemistry as well, from the University of Pennsylvania.

Again, and it may be something of a surprise to those who knew him—he was a commissioned officer in the U.S. Army for a period of six years. His active duty as a second lieutenant was first as a cavalry officer and later as a field artillery officer. Kirk's mother had a strong desire that he should become a "Christian soldier."

He received his PhD in biochemistry from the University of California at Berkeley in 1927, and was appointed a professor of biochemistry in 1929.

He soon developed a reputation as a microchemist. This was before the days of sensitive analytical instrumentation, and biochemical analysis hinged on scaling down classical chemical techniques to deal with microgram and submicrogram quantities of materials. He developed a whole series of glassware for various types of analysis and invented the quartz microbalance. His first exposure to a physical evidence case was in 1935, and that experience, by all appearances, was a gratifying one. He began to work more and more in the physical evidence area.

He was sidetracked by World War II, however. Because of his adeptness in dealing with microsamples, he was asked by the Manhattan Project to undertake the

---

[1] Reprinted by permission of John I. Thornton from a speech given before the California Association of Criminalists Founder's Seminar May 1989.

first microscale production of plutonium. The first reduction of plutonium salt to plutonium metal in 1942 represented only about 15 micrograms. More than 500 tons of plutonium has since been produced.

After the war, he returned to teaching and research and gradually weaned himself off biochemistry and into criminalistics. In 1954 he formally disengaged himself from the Department of Biochemistry and became affiliated with the newly formed School of Criminology at UC Berkeley. He was a professor of Criminalistics until his retirement in 1967, but as an emeritus professor did in fact retain his academic citizenship until his death in the summer of 1970.

During his career he wrote over 250 articles, monographs, and four major texts. At the time of his death he had worked on 2200 criminal and civil cases.

It's useful to look at a few events in Dr. Kirk's career to see how those events shaped his outlook, and to see how his outlook has influenced all of us.

The case that vaulted Dr. Kirk into regional and even national prominence was the Burton Abbot case. In April of 1955, Stephanie Bryant, a young teenage girl, disappeared from the tennis courts at the Claremont Hotel, at the Berkeley/Oakland line. An intense search was made for her, and public awareness of the case grew exponentially. But then in July 1956, Coorgia Abbot found Stephanie Bryant's purse in the basement of her house. Burton Abbot was the other resident. Abbot denied all knowledge of the purse (the basement was kept unlocked), and denied having ever seen Stephanie Bryant. But further search of the basement revealed, buried, Stephanie's eyeglasses, and other aspects of the case.

At this point, the physical evidence isn't really the story. The story is the incredibly intense publicity that the case received. Abbot was brought to trial, the trial lasted 12 weeks, which was a very long trial in those days. Abbot was sentenced to death, and was executed within a year.

Of the 114 days pretrial, there were only 16 days in which a story on the case was not carried by either the San Francisco Examiner, the San Francisco Chronicle, or the Oakland Tribune. Even prior to trial, the case occupied the front page of the Oakland Tribune for 37 days straight.

All of this was only about a year after the publication of Dr. Kirk's textbook, *Crime Investigation,* in which he attempted to establish a discipline and indeed a profession devoted to the examination of physical evidence. His role in the Burton Abbot case was indeed a fulfillment of all of those things he had been trying to say in his textbook. To the lay public, his involvement in the case shifted the case from just a good circumstantial evidence case to one that would convince anyone, even the most skeptical of people. But to the scientific community (and to the legal community, but for different reasons), Dr. Kirk's role in the Abbot case did a great deal to establish scientific legitimacy to the criminalistics enterprise. With the scientific community, the Abbot case gave Dr. Kirk a pulpit, and he used it vigorously and effectively.

Listen to his comments to the American Chemical Society, "If you went home some night to find your wife murdered, few of you would attempt to solve the crime yourselves. With all due respect for your confidence in your professional colleagues, I do not believe you would call another analytical chemist to solve the crime for you either. You would call the police. Why then would you expect the police to turn around and give the matter back to another chemist?"

Chemists (and others) listened to that, and it made sense to them. And that probably marked an epoch in the history of our profession. Other responsible professions—chemistry, medicine, law, and the police—acknowledged that this was a serious scientific endeavor, and not just high-grade detective work.

Following the Abbot case, there were a great many people who knew the name Paul Leland Kirk, and even with those who didn't know his name, there were quite a number who knew that there was "some professor" at Berkeley who was adept at physical evidence matters.

The other principal case was, of course, the Sheppard case. To my way of thinking the Sheppard case wasn't as pivotal to our profession as the Abbot case, but it did keep Dr. Kirk in the spotlight and again, he came off on the winning side. The Sheppard case actually predated the Abbot case by a few months. Following the conviction of Dr. Sheppard at the time of the first trial, Dr. Kirk examined the Sheppard crime scene in January of 1955, spending four days at the scene at the request of the defense. His report on the physical evidence became known as the "Kirk Affidavit" when it was appended to a motion for a new trial. This affidavit comprised 33 typewritten pages and incorporated by reference 16 supplemental pages as appendixes and was accompanied by 46 photographs taken by Dr. Kirk himself. The affidavit presented evidence which could only be interpreted as pointing the finger of guilt away from Sam Sheppard and toward someone who was left-handed. The prosecution had some snotty things to say about Dr. Kirk's affidavit, in general, but it was never refuted. The affidavit of course did not get Dr. Sheppard a new trial. What got him a new trial was the adverse publicity at the time of the first trial, which led the U.S. Supreme Court to conclude that the trial judge did not fulfill his duty to protect Sheppard from inherently prejudicial publicity which saturated the country.

The second trial in 1964 lasted only three weeks: the first trial had lasted two months. Dr. Kirk testified and the prosecution witnesses did a lot of tap dancing and backpedaling. And one prosecution witness conceded on cross-examination that she was now convinced that the assailant was left-handed. The jury acquitted Dr. Sheppard in just a few hours of deliberation. This was, of course, a vindication of Dr. Kirk as well.

In discussing the Abbot and Sheppard cases I've focused on publicity. Certainly that had a role in the Paul Kirk story, but it wasn't the only thing. Let me try to identify some other factors.

Before he became interested in criminalistics, he was established as a good scientist. Because of his background, he saw criminalistics as a science discipline, which was virtuous in its own right, not just an extension of police work. And as a corollary to this, he recognized that many case situations are actually *research* problems, calling for the application of the scientific methods. He was a brilliant man. Now he was human, and he made mistakes like all of the rest of us. He wasn't infallible, he was vain in some respects, but he was honorable, and he was possessed of a truly first-rate intellect.

He touched all of us personally—everyone who knew him will have his or her own story about their interaction with him.

He wrote a lot, and influenced people by his writing. Not just people in our profession, but people in the legal profession as well as the police, the judiciary, and other scientific disciplines. Well, he had to write a lot because he was situated in a research university where publications were held to be important. Had he been at some other institution even with the same gifts and the same interests, he may not have written as much—the pressure just wouldn't have been there.

He was truly a professor—he had something to "profess."

And he loved the profession.

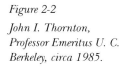

Figure 2-2

John I. Thornton, Professor Emeritus U. C. Berkeley, circa 1985.

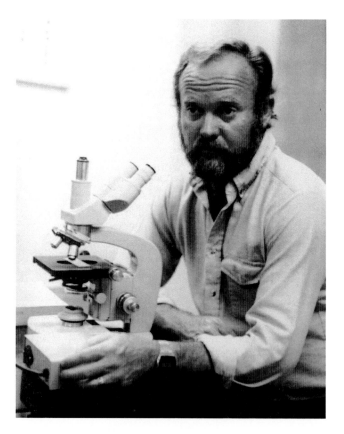

Figure 3-1   Representatives of Newtonian and non-Newtonian pressure streaming is shown. From Rheological Phenomena in Focus, *by David Boger and Kenneth Walters. (1993). Elsevier, Amsterdam. Reprinted by permission of Dr. David V. Boger.*

CHAPTER 3

UNDERSTANDING BLOOD BEHAVIOR

WHAT DIFFERENCE DOES NON-NEWTONIAN FLUID BEHAVIOR MAKE?

Blood Dynamics alerted bloodstain pattern analysts to the fact that regarding blood as a non-Newtonian[1] fluid could be significant in explanations of behavior.[2] This approach to understanding blood had not been evaluated in previous literature, where Newtonian fluid behavior traditionally was accepted as the guiding principles. Newtonian fluids may be described as liquid drops that form by virtue of surface tension and may oscillate during flight away from their source. Non-Newtonian substances separate into drops by virtue of internal cohesion, and may demonstrate a remarkable stability in flight. Blood can actually behave as either Newtonian or non-Newtonian, depending upon conditions of the victim and location of the injury. A misunderstanding, however, has arisen from interpreting non-Newtonian behavior as not following Newton's Laws of Motion. Non-Newtonian behavior has nothing to do with the physics Laws of Motion established by Sir Isaac Newton.

Blood is now classified as non-Newtonian,[3] in contrast to other aqueous (watery) fluids which are called Newtonian. The latter (Newtonian) defines behavior for water, ink, and blood serum, while the former (non-Newtonian) defines liquid plastics, mud, food suspensions, paints, red blood cells, and other colloids and complex substances, in terms of friction to flow. Following the notation of blood's classification, statements were made that translated "non-Newtonian" behavior into "not following Newton's Laws of Motion." This was an error in translation.

Sir Isaac Newton died in 1727 after considerable contributions in proposing Laws of Motion and defining gravitation within the then-emerging science of physics. In 1840, fluid flow was defined by a formula involving friction-based principles, suggested some 120 years earlier by Newton.[4] This behavior was named Newtonian fluid in Newton's honor, but was not related to his Laws of

---

[1]Of the many spellings available, the form used in *Journal of Non-Newtonian Fluid Mechanics,* McKinley, G.H. and Keunings, R, Eds., Elsevier, Massachusetts, will be used here.
[2]Wonder, *Blood Dynamics*, 9.
[3]Nubar, Yves. (1966). *The Laminar Flow of a Composite Fluid, an Approach to the Rheology of Blood.* New York Academy of Science, New York, 35.
[4]*Ibid.* [Nubar], *Rheology of Blood.*

Motion. The principles used to define Newtonian fluids were taken from his *Principia*, which said:

> The resistance which arises from the lack of slipperiness of the parts of the liquid, other things being equal, is proportional to the velocity with which the parts of the liquid are separated from one another.[5]

Slipperiness was later labeled as viscosity. One hundred years later, in the 1940s (213 years after Newton's death), it was found that not all fluids behaved according to the formula postulated from Newton's suggestions. These other observations followed development and manufacture of synthetic materials, following World War II, and the industrialization of the world, with substances such as nylon, plastics, petroleum byproducts, and commercial mass production food stuffs (fluids unknown during Newton's life). Since the fluids didn't behave according to formula derived for Newtonian ones, they were then called *non*-Newtonian. These then required new formula and study regarding their behavior.

Some studies were already done under different names, beginning with water clocks (Figure 3-2), but were gathered and redefined around 1926 under the title of rheology.[6] Hydrodynamics is the study of the flow of water and other Newtonian fluids, while rheology is the study of the flow of non-Newtonian fluids including blood, when it flows as a non-Newtonian fluid.

Volume (pool): An accumulation of blood

Various attitudes have been expressed toward incorporating non-Newtonian fluid behavior into bloodstain pattern analysis. One viewpoint is to ignore it or claim that it applies only to flowing blood, such as **volume (pool)** stains spreading out from a bleeding victim. Although this is a possible application of the fluid mechanics, it is a limitation to the usefulness for the information. With non-Newtonian fluids, boundary layer flow refers to many combinations such as a fluid flowing within a fluid (oil in water), gas within a gas (poisonous gas within air), and solids within solids (rock strata in mountains), and all permutations such as a liquid flowing within a gas (blood drop behavior through air) and solids within a liquid (blood cells within plasma, which contributes to the dynamics of drop separation). The contrast between principles of both Newtonian and non-Newtonian behavior applies to the scientific understanding of how blood drops are distributed during many dynamic events. Blood flows as either Newtonian or non-Newtonian depending upon circumstances, and these can relate to crime events.

Bloodstain pattern training: A technique to determine cut off points on a measured blood spatter.

Because this information is new to traditional **bloodstain pattern training**, and sometimes at odds with traditional definitions of blood behavior, there may be

---

[5]Barnes, H.A., Hutton, J.F., and Walters, K. (1989). *Rheological Series 3, An Introduction to Rheology.* Elsevier, Oxford, 1.
[6]Reiner, Marcus. (1964). The Deborah Number, *Physics Today*, January, 62.

resistance to incorporating dramatically new information. Accurate principles, however, are essential for advancing applications of techniques which will benefit analysis and subsequent investigations utilizing the evidence. Since modern crime labs are increasingly staffed with individuals entering from firm academic backgrounds, there should be acknowledgment for technical background material in understanding blood behavior during the dynamics of criminal violence. Hopefully, science updates will replace pre-1940s' beliefs. Descriptions of blood behavior in terms of viscosity, surface tension, and proportionality formula will be replaced with understanding non-Newtonian blood in terms of cohesion, hematocrit influence,[7] and vascular injury dynamics upon blood drop separation and distribution.

To avoid confusion since blood, under some conditions, can behave Newtonian, when Newtonian behavior is noted the phrase will be "drop formation" and with non-Newtonian behavior the phrase will be "drop separation." Blood drops form by Newtonian principles depending upon surface tension and behave as viscous substances. Blood drops separate as non-Newtonian substances from their blood source depending upon cohesion, which is influenced by the ratio of red blood cells, freshness of the blood, and changes such as coagulation. These both can be the subject of future research and further definition.

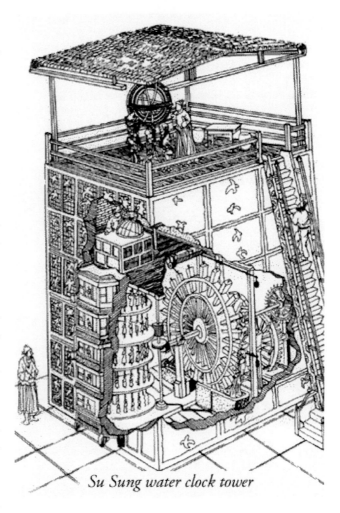

Su Sung water clock tower

Figure 3-2

Early water clock designs were later defined as Newtonian behavior regarding simple flow based on suggestions by Isaac Newton.

Although the study of Newtonian fluids, with water as the representative, has a long and well-studied history starting with climate and weather research, the study of non-Newtonian fluids is relatively new in science. In fact, Dr. Paul Kirk was most likely unaware of the classification, since his basic academic studies were completed prior to 1940. A lot more than just the liquid flow is involved, yet the extent of involvement has not been fully understood up through 2006. Examining what actually occurs during distribution at criminal events requires

[7]*Erythrocyte Mechanics and Blood Flow.* (1980). Giles R. Cokelet, Herbert J. Meiselman, and Donald E. Brooks, Eds. Allan R. Liss, Inc., New York, 75.

consideration to the parts as they relate to non-Newtonian behavior. The parts applicable to BPE suggested at this point in time:

- Formula and units of measure comparison with Newtonian fluid mechanics
- Axial flow as it relates to Newtonian and non-Newtonian fluids[8]
- Affect of blood composition on classification as non-Newtonian
- Crime events that are affected by fluid mechanics
- Applications which one day may benefit BPE

FORMULA AND UNITS OF MEASURE COMPARISON WITH NEWTONIAN FLUID MECHANICS

The formulas for the two classes of fluid mechanics provide different units of measure.

The Newtonian formula (resulting from Newton's suggestions)[9] is:

$$\tau = \mu \, dv/dy$$

τ (Greek letter Tau) equals μ (Greek letter Mu) times dv/dy (Calculus differential) where:

τ = shearing or frictional stress per unit area against the walls of the containing vessel or the enveloping air around a drop, called a **Reynold's number**

μ = coefficient of viscosity, a *constant* of proportionality

dv/dy = a linear change in velocity gradient of flow

Essentially Newtonian flow is dependent upon a viscosity constant and surface tension, where each is proportional to the rate of flow. Most important, each can be measured and stated with regard to a stationary quantity of fluid then applied to the flowing liquid. The unit of measure is the Reynold's number. The higher the viscosity and/or flow rate, the higher the Reynold's number.

Later it was found that non-Newtonian fluids follow the formula[10]

$$\tau = \kappa \, (dv/dy)^n$$

where κ (Greek letter Kappa) is a constant but not of viscosity, since viscosity is no longer constant, thus the constant designation Kappa rather than Mu.

---

[8]*Blood Flow Theory and Practice.* (1983). Taylor, D.E.M., Ed. Academic Press, London, 288.
[9]Vennard, John K. and Street, Robert L. (1982). *Elementary Fluid Mechanics*, 6e. John Wiley and Sons. New York, 14.
[10]*Ibid*, 16.

Although the shear rate (friction) is constant at any given velocity, it changes depending upon flow rate, thus now equals a changing velocity gradient. The change with Newtonian fluids is linear but exponential with non-Newtonian fluids. The unit of measure is no longer a Reynold's number since it is describing an incomparable behavior. For non-Newtonian fluids the shear or frictional stress, Tau, is called a **Deborah number**.

The term Deborah was derived from the Biblical quote of Deborah who says in the Bible,[11] "everything flows before the Lord." The definition for units in rheology followed from the concept that "everything flows if you give it enough time." Some "fluids," in addition to the plastics, defined as non-Newtonian early in the discovery were not those commonly thought of as fluid i.e., rocks and glass. Non-Newtonian substances, first considered, flowed slower than Newtonian because of the increasing shear (friction) against the path they flowed. The formulas themselves are of interest to scientists conducting academic research, industrial engineers dealing with non-Newtonian fluids in production, and geologists. From a forensic science level the comparison between the conditions when blood behaves non-Newtonian compared to the conditions when blood behaves Newtonian may be more important than the formulas. This is because both kinds of flow may occur, separated, in the body. Which blood vessels demonstrate which behavior at any given time is determined by hormone systems.

Learning to recognize characteristics of each type of flow within bloodstain patterns can provide scientific evidence for locating the injury responsible for the bloodstain. Under stress trauma to the body, not necessarily an emotional reaction; the rate of flow may shift. If excess rates of flow are required for preservation of life (i.e., in the event of massive blood loss and/or excess adrenalin reaction (fight or flight)), the body may momentarily change non-Newtonian flow to Newtonian by uptake or extrusion of body fluids. The balance of fluids and cells in blood decreases or increases the red cell ratio, hematocrit. It is primarily the red blood cell ratio that determines whether blood will flow as Newtonian or non-Newtonian. Bloodstain patterns can reflect this shift in behavior.[12]

Part of the complexity in defining flow with non-Newtonian fluid stems from variable behavior that involves three types of flow rates instead of one constant proportionality as with Newtonian fluids:

---

[11]Judges 5:5. Most Bibles list the quote as Deborah saying "the mountains melted before the Lord." This was explained by the Israeli physicist and research professor of the Israeli Institute of Technology, Marcus Reiner, at the fourth International Congress on Rheology in Providence, RI. (1963). English translators changed the original Hebrew from "the mountains flowed before the Lord" to "melted." The quote used here is translated from the original Hebrew.

[12]Kim, Sangho, Young, I., Cho, Jeon, Abraham, H., Hogenhauer, Bill, and Kensey, Kenneth R. (2000). A new method for blood viscosity measurement. *Journal Non-Newtonian Fluid Mechanics,* Elsevier, Amsterdam, Vol. 94, 55.

1. Where shear (boundary/interface friction) increases with the speed of flow (shear thickening fluids where the exponential factor is n > 1).

2. Where shear (boundary/interface friction) decreases with speed of flow (shear thinning fluids where the exponential is n < 1).

3. Where a shift between the two occurs at higher velocities of flow. This happens as shear thickens to a point then reverses and thins at still higher velocities. This latter is called *thixotrophy*, and blood flow within vessels is thixotrophic. This contributes to vascular problems with high blood pressure.

What must be understood as this applies to blood is that surface tension and viscosity cannot be measured on stationary fluid. Values of surface tension and viscosity obtained with the same techniques and formula as for Newtonian fluids but applied to non-Newtonian fluids are completely invalid.[13] The values change depending upon the flow rate, i.e., blood must be flowing when measured for surface tension and viscosity in the study of rheology.

AXIAL FLOW AS IT RELATES TO NEWTONIAN AND NON-NEWTONIAN FLUIDS

It was pointed out in *Blood Dynamics*[14] that blood does not flow evenly mixed within blood vessels. Flow is normally a core of red blood cells with a layer of white blood cells around the core and circulating plasma with platelets scattered within the fluid enveloping the core. This is non-Newtonian flow for normal concentrations of red blood cells. The cohesion created by the proportion of red cells causes the non-Newtonian behavior, not the fact that blood appears viscous.[15] This type of flow is stable at normal flow rates, but cannot sustain increased flow without destabilizing. When the column destabilizes, axial flow is disrupted. To sustain high rate of flow (i.e., high blood pressure), the body shifts fluid from the surrounding tissue into the blood vessels, noted as thirst for victims with high blood pressure, and those suffering shock. The significant fact from the bloodstain pattern is that the increase in aqueous (water) fluid within a blood vessel may lead to shifts in fluid behavior from non-Newtonian behavior to Newtonian.

Newtonian fluid can sustain the higher velocities of flow but is essentially destabilized; this is called turbulent flow. In turbulent flow, red cells are no longer flowing as an axial core but scattered through the fluid. When these cells come into contact with blood vessel walls they may stick, thus creating lumps and breaks in the normal flow of fluid. This leads to narrowing of vessels and vascular diseases including heart trouble and stroke. An interesting note

---

[13]*Ibid*, 35.
[14]Wonder, *Blood Dynamics*, 10.
[15]Nubar, *Rheology of Blood*, 35.

is that in the laboratory one can change some non-Newtonian chemicals by adding other chemicals such as glucose (sugar). If the same happens in living blood vessels, it might explain why diabetics are more likely to die from vascular disorders than from the mere presence of high levels of blood sugar.

EFFECT OF BLOOD COMPOSITION ON CLASSIFICATION AS NON-NEWTONIAN

The ratio of red blood cells in blood determines whether the fluid will behave Newtonian or non-Newtonian because plasma is usually considered Newtonian. If red cells are removed or low due to health conditions, such as in anemic people, normally non-Newtonian flow may shift to Newtonian fluid behavior. Non-Newtonian blood drop separation has been defined. This is seen in the popular film from the British Home Office London Laboratory, *Blood in Slow Motion*.[16] The fluid column goes through the following changes for drop separation (Figure 3-3):

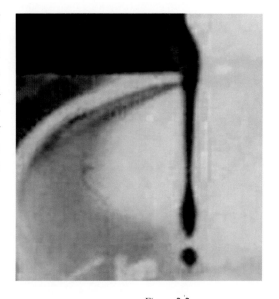

- Extension (also called elongation)
- Truncation (also called bifurcation, i.e., narrowing of the column sides; depending upon elasticity)
- Separation into drops (insufficient elasticity)
- Retraction of fluid back into the column (sufficient elasticity)

Figure 3-3

Illustration of blood drop separation as determined in the science of rheology with stages of extension (elongation), truncation (bifurcation), retraction, or separation.

None of this considers surface tension in the process for two reasons: surface tension is an insignificant force in comparison with internal cohesion, and surface tension varies constantly and cannot be measured by normal methods. Newtonian fluids, such as water, depend upon surface tension to determine drop formation because surface tension is all that applies. There is also no such thing as cohesion of surface tension. Cohesion (internal fluid attraction), adhesion (fluid attraction to a surface), and surface tension (fluid surface forces) are all separate entities.[17,18]

CRIME EVENTS WHICH ARE AFFECTED BY FLUID MECHANICS

Blood inside the body behaves much the same way as that of blood outside the body, with the exception of drying. The fact that bloodstains dry outside the body serves as a record of which events happened at the time of and following bloodshed. Determining what happened inside the body is much more complicated

---

[16]*Blood in Slow Motion* (Video). (1991). Home Office Main Laboratory, London.

[17]Epstein, Lewis Carroll. (1993). *Thinking Physics, 2e*. Insight Press, San Francisco, 191.

[18]*Concise Encyclopedia of Science & Technology*. (1998). McGraw-Hill, New York, 25, 435, 1930.

and requires instrumentation of a clinical laboratory. On the other hand, research into blood behavior in the body provides considerable information, which can be used to understand blood behavior outside the body.

While non-Newtonian behavior is the primary classification throughout most of the blood vessels, a few vessels require Newtonian flow in serving organs. Three of these are the brain and adrenal glands, supplied through the carotid arteries, and the heart, supplied from pulmonary vessels with output through cardiac arteries. This becomes essential when challenge to these organs occurs. The body reroutes 85 to 90 percent of all blood flow to them,[19] thus concentrating the blood supply. The need for increased flow with increased volume input requires non-Newtonian to shift to Newtonian in time of need by the heart and brain. During this time cutting the carotid arteries of the throat can project a blood column of destabilized Newtonian flow. An example of Newtonian versus non-Newtonian flow is shown in Figure 3-1. The projected pattern for Newtonian flow can be a central dense circle with small to fine satellite spatter around it. Since this may be specific for carotid arterial damage, it locates the event as well as correlating bloodstain pattern with the appropriate injury to the victim.

It has been suggested that the central core with satellite spatter type of pattern results from the high pressure of the arterial blood column striking a surface. The dynamics of this would more likely project satellites back, away from the target or outward as streaks as in Figure 3-4 (demonstrated with aging blood behaving Newtonian), rather than around the circumference. The destabilized Newtonian behavior, by contrast, would record the less dense spatter ring around the solid core as small dots not streaks.

Another crime scene application of non-Newtonian versus Newtonian may involve the situation of coagulation. When blood is fresh and pools rapidly (forms a volume stain), the internal cohesion creates a strong bond holding the pool together. As the blood clots and retracts, the red cells are pulled into a core with serum extruded, pushed out. Serum is Newtonian, lacking the internal cohesion of the red cells, thus will flow off the **clot** by the effects of gravity. At a crime scene the location of the retracted clot would be more important than the flow of serum in determining where the bloodshed volume accumulated (pooled).

Clot: A physiological change in blood after it leaves the body.

Figure 3-4

Ricochet model of satellite (secondary) drop distribution. Blood used was out of date whole blood diluted with broth and pressurized. Behavior would be Newtonian.

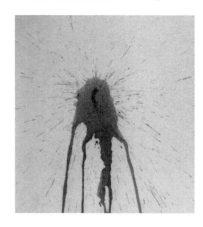

APPLICATIONS THAT ONE DAY MAY BENEFIT BLOODSTAIN PATTERN EVIDENCE

Someday applications may be found for properties of non-Newtonian blood. In a laboratory, whole blood, blood with a ratio of red blood cells

---

[19]Sohmer, Paul R., MD. (1979). The Pathophysiology of Hemorrhagic Shock. In *Hemotherapy in Trauma & Surgery*, AABB, Washington, D.C., 2.

to plasma at around 45 percent or higher, may be seen in a capillary pipette as a convex meniscus. Water and other Newtonian fluids show an inverted meniscus (i.e., concave). The reason for the difference is in the response of each type of fluid to the walls of the capillary and whether surface tension or cohesion governs behavior. This is a demonstration of the difference between surface tension and cohesion (Figure 3-5).

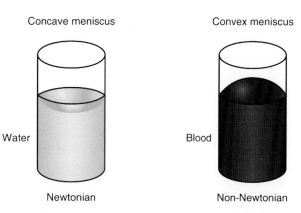

Figure 3-5

Meniscus difference between Newtonian and non-Newtonian. This is also a comparison of surface tension versus cohesion.

Because of the properties influencing curvature of the fluid surface, non-Newtonian behavior can provide a contrast in reflected light. Newtonian fluids, governed by surface tension, show straight edges except for curvature at the edge of a volume (pool). Non-Newtonian fluids, on the other hand, show a much greater distortion to a reflected image (Figure 3-6). This property could be compared to fluid reflections such as water (rainfall), oil and automotive fluids, and volume (pool) bloodstains at the scenes of traffic accidents on black top, or where only black and white crime scene photographs are available later.

Differences in behavior when stirred are also noted. At this time many of the facts known regarding non-Newtonian fluid behavior, and perhaps more important the contrast between different types of flow inside and outside a body, may seem unimportant to crime investigation. It should be remembered, however, that DNA was once thought to be insignificant to crime solving too. We do not have all the answers now but the importance is that bloodstain pattern analysts be aware and some be involved in looking for future applications of the information.

Figure 3-6

Reflected image differences with non-Newtonian versus Newtonian fluids.

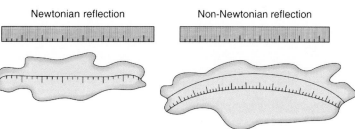

Figure 4-1   Not all people hear things the same way.

CHAPTER 4

TRIGONOMETRY IN BLOODSTAIN PATTERN EVIDENCE, MATH USE IN QUESTION

TRADITIONAL DERIVATION OF FORMULA

Traditionally the application of trigonometry to determine the origin of an impact event that distributed blood drops has been based on a series of theorems and assumptions:

1. A spherical drop makes an elliptical cross-sectional at contact with a surface (Figure 4-2).
2. A line intersecting parallel lines will create equal angles.
3. The diameter of a blood drop is equal to the width of the projected ellipse.
4. The length of the ellipse can be determined by using the shape of the blood drop's first contact with the surface and completing the ellipse shape as indicated.
5. One can construct a right triangle from contact lines drawn from the theoretical blood drop to the approximated ellipse (Figure 4-3).
6. Measuring the width of a stain and dividing by the length provides a number of less than one (<1), which can be used with a pocket calculator or log table to find the arc sin that is the angle of incident for the original blood drop.
7. Using several measured stains, one can backtrack from the incident angles and locate an area containing the origin of a blood distributing impact event.

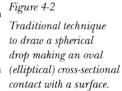

Figure 4-2

Traditional technique to draw a spherical drop making an oval (elliptical) cross-sectional contact with a surface.

A fundamental problem with this model occasionally has been pointed out by students. A drop of blood cannot have the same cross-section measurement as the bloodspatter it leaves. This is because the drop is three-dimensional whereas the stain is essentially two-dimensional, with conservation of volume spread over a wider area. The bloodstain must always appear bigger than the drop (Figure 4-4). A right triangle can still be constructed based upon the resultant bloodstain, but the angles involved do not include a tangent (the length of the stain and hypot-

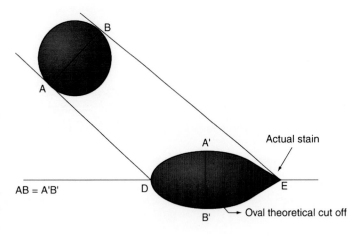

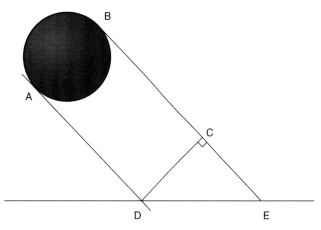

Figure 4-3

Rotation of theoretical contact lines to conform to a trigonometry model.

Figure 4-4

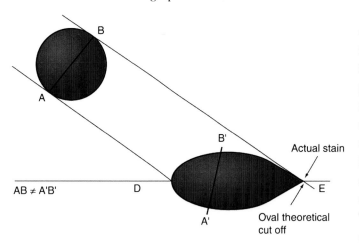

enuse of the triangle) through parallel lines, thus does not involve equal angles, internal and external. The reason for constructing the right triangle is to determine the incident, first contact angle that is an external (outside the right triangle) angle. Only if the external is equal to the internal can we calculate the incident angle by measuring the stain.

Most instructors answer questions of disparity in size by claiming that proportionality remains the same. This may justify the process but no specific study has been found that shows that there really is proportionality. Advocates of the technique claim that since the ellipse is completed based on the initial contact angle location, the angle would be the same at the lower end of the duplicated curvature. This leads to the next problem.

The first contact between a blood drop and a surface is not, in fact, an elliptical contact. When a spherical blood drop first contacts a surface, it is from collapse of a chord (pronounced *cord*) of the sphere. Watching blood drops make narrow angled contact from the side implies contact is a tear drop shape when, in fact, contact is a series of cross-sections of the sphere (i.e., circles; Figure 4-5). Even more significant is that the diameter of the chord, contact is dependent upon the velocity of the drop, which follows Paul Kirk's logic of stain shape affected by drop velocity. This can be seen in published photographic studies.[1]

Unfortunately drop studies predominantly have used blood dripping from droppers onto slanted targets. Drips include large drops in free fall due to gravity alone. References have stated that the shape of the ellipse depends upon the angle with which the blood drop meets the target surface.[2] That may be of lesser importance than Dr. Kirk's original observation that the velocity of the blood drop affects the shape of the final stain. The faster the blood drop is traveling at contact with the target, the less initial

[1]Bevel, Tom and Gardner, Ross M. (2002). CRC. Boca Raton, FL, 133.
[2]James, Stuart H., Kish, Paul E., and Sutton, T. Paulette. (2005). *Principles of Bloodstain Pattern Analysis Theory and Practice*. CRC, Boca Raton, FL, 221.

footprint the sphere will make. The slower a drop is traveling at initial contact, the wider diameter chord of the sphere will be initially recorded. This is seen even with the variation of blood dripping onto angled cardboard. The wider the stain, the more open the angle because the ratio is larger. The faster a drop travels until all forward momentum is dissipated will lengthen a bloodstain as well as narrow it to produce an apparently smaller degree angle of contact. Gunshot Distributed Impact Spatters (GDIS) produce streaks when recorded even though they may actually be found at what should have been a calculated angle to the origin.

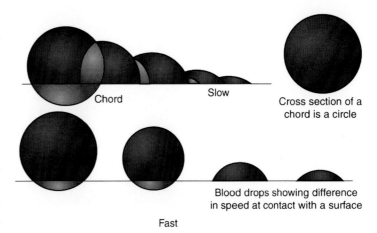

Figure 4-5

Chord (pronounced "cord") of a sphere deposited, depending on speed of the drop.

The theory of terminal velocity was introduced as a means to achieve a known, or at least constant, velocity for blood drops. Blood was dripped onto slanted targets from greater heights in an effort to create blood drops at known velocity striking a surface at known angles. Terminal velocity applies if an object falls beyond a certain distance. The object will accelerate according to gravity until that point beyond which acceleration is in equilibrium with wind shear and other forces. Falls continuing after equilibrium will be at uniform velocity. Unfortunately, the terminal velocity point depends upon the mass of the free-falling object in air. Only in a vacuum can mass be ignored. In air, since blood drop composition and mass are not uniform, nor can they be known for any given blood drop distributed during a violent crime, terminal velocity will vary. The major problem with introducing terminal velocity, however, is that it does not deal with blood drops projected by force. The only application is for drip cast offs (LVIS, passive stains, drip trail, gravitational drops).

Assumptions have been made that the variance is small in blood drop mass. This is based on the false concept[3] that blood is uniform in red blood cell ratio (hematocrit) or reasonably so over the normal range. However, blood distribution is not limited to that from individuals in normal health. Anemic, malnourished homeless, alcoholics, drug addicts, elderly, children, aborted fetuses, and people suffering heart attacks all may be considerably outside the normal ranges for red blood cell ratios, and these are mentioned frequently in victimization studies.[4] In addition, it was pointed out in *Blood Dynamics*[5] and included in the

---

[3]Albert, Soloman N. (1971). *Blood Volume and Extracellular Fluid Volume, 2e*. Charles C Thomas, Springfield, 50f.
[4]MacDonald, W. F., Ed. (1976). *Criminal Justice and the Victim, Vol. 6*. Sage, London, 9–279.
[5]Wonder, *Blood Dynamics*, 8.

Solid Object

**Projection Application
Transference logic**

Shadow

Figure 4-6

The projection of a shadow creates an oval shape from a spherical ball.

Figure 4-7

Shadow of the suspended ball looks good, and shows contacts of parallel lines drawn to an oval (ellipse).

appendix here that red blood cell ratios vary greatly for different organs of the body. The assumption that blood distributed during violent crime is always of normal hematocrit ratios of red cells is unrealistic.

Attempts to locate the mathematical justification from the original logic have been unsuccessful. The closest approximation could be the projection of a sphere as a shadow (Figure 4-6). The elliptical shadow, seems to lend itself to drawing of contact points from the ball to the length and width of the shadow (Figure 4-7). The angle estimation, however, is not a function of contact between the sphere and a surface, but one dependent upon the positioning of a light source (Figure 4-8). Although the shadow is projected from a sphere, a flat circle would produce the same affect. This is not the same dynamics as a three-dimensional liquid drop contacting a surface at a specific angle to produce a two-dimensional elliptical stain. A projection image formula should not be applied to a transfer of substance.

An exercise that illustrates the impossibility of creating an oval cross section from a sphere is to take a spherical object and attempt to make a single, straight, diagonal cut across it, producing an oval or ellipse. The result will actually always be a chord of the sphere with a circle as the cross-sectional shape. No matter how you slice it, any cross-section of a sphere will be a circle, not an oval (ellipse). It is physically impossible and contrary to geometry to cut a sphere so that an oval cross-section results (Figure 4-9).

Two questions follow this logic. Why do analysts still use the trigonometry derivations, and why haven't researchers noticed the lack of logic behind the traditional theories? The probable answer is because the technique seems to work when applied to mock crime scenes, and mock scenes have been constructed predominantly with blunt force impacts instead of firearms, two quite different degrees in velocities. To test the answer regarding comparative velocities, a series of experiments was conducted at the California Highway Patrol Academy firing range in Sacramento on 15 July 2005.

PRELIMINARY STUDY OF MEASUREMENT TECHNIQUES AND BLOOD DROP VELOCITY EFFECTS

In July of 2005, an experiment was conducted at the California Highway Patrol (CHP) Academy near Sacramento, California. The class consisted of 15 officers of the CHP, none of whom were previously trained in bloodstain pattern analysis. The scenes were to be seven rooms at the firing range with a dirt berm back drop. Six rooms were used as gunshot experiments. The seventh was used to construct a beating assault mock scene for comparison. Three firearms were available: a 1930s' issue S&W .32 semiautomatic pistol, a .22 revolver (manufacturer unknown), and a recent issue S&W .40 with hollow point rounds. The first scene set with the .22 had to be eliminated because of lack of bloodspatters recorded on a wall adjacent to the shot. The .32 produced some bloodspatter, but no measurable stains were found after screening the wall with a magnifying page sheet reader; only streaks and dots were noted. Of the remaining four gunshot experiments, the .40 caliber was used but one other scene was considered unsatisfactory and eliminated.

Human packed red blood cells with a high hematocrit of 89 percent were used fresh, not yet outdated, in aliquots of 5 cubic centimeters in plastic sandwich bags. The seventh scene was constructed by hitting an exposed 5 cubic centimeter volume (pool) of blood with a hammer.

The beating scene was constructed by placing a 5 cc blood volume (pool) on an up-ended wastepaper basket and striking once with an overhand blow of a common carpenter's hammer. The basket was removed and hidden from view of the participants.

The group was shown two methods to measure spatters. Time constraints and weather conditions prevented prior set up of the experiment, therefore original plans as a double-blind ended as a single-blind exercise. Three instructors knew the actual positions of the origin, and two did not. The two measuring techniques were labeled oval, for completing the ellipse of each spatter (Figure 4-10) and BPT, for the bloodstain

Figure 4-8

The problem with projection formula is that the shadow is a function of the light source, not a relationship between a sphere and its contact with a surface.

Figure 4-9

No matter how you slice a sphere, a circle cross-section will result.

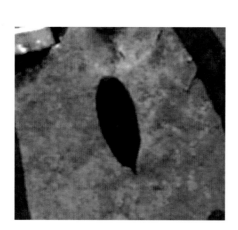 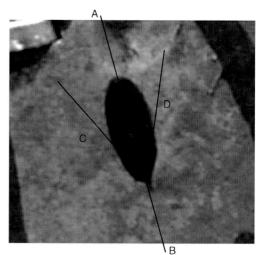

Figure 4-10

Complete the Oval:
A method of measurement involves completing the oval of a spatter. The edge of first contact is used to approximate the opposite edge, and the oval shape is measured.

Figure 4-11

BPT. Another method is taught in some work-shops. Participants are trained where to cut off the stain with some of the overflow but not all. A modification of the technique is to use the sides of the tear drop and draw lines that meet at an extension of the overflow.

Areas of convergence (AC):
An area determined from the directions of travel of several impact distributed spatters.

pattern training technique used in 40-hour workshops of at least four different instructors (Figure 4-11).

The final arrangements of the experiment were severely limited in time and participation; therefore, changes were made to involve four scenes; three were gunshots and one was a beating with only one technique in measuring for each. Some observations, however, could provide information for future research. Positions used in the experiments are shown in Figures 4-12 through 4-17 (see pages 39–43).

On the basis of this experiment, it could not be resolved as to which measuring technique was better. Observations during the experiments and previous experience with the techniques viewed during the two-day Advanced Blood-stain Pattern workshop conducted for the American Academy of Forensic Sciences (AAFS) in Atlanta, Georgia (2002) did indicate that completing the oval, especially when used with the concept of point of origin, can result in considerable error for final interpretation. Completing the oval technique cuts the stain off too soon and results in a more open angle than the reconstruction requires. Drawing the angle back to a point instead of a general area may shift the origin well beyond the true location.[6] Either technique, oval cut-off or drawing stains to a point origin, used alone results in less error than when they are used together. Fortunately newer computer applications are directed toward **areas of convergence (AC)** (or origin) over points of convergence (or origin).[7]

---

[6]Wonder, *Blood Dynamics*, 44–45.
[7]Carter, A.L. (2005). Further Validation of the BackTrack™ Computer Program for Bloodstain Pattern Analysis. *IABPA News*, **21**, 3, 15.

Surprising, however, was that the oval technique seemed to work better for more rounded stains such as those associated with 70- to 90-degree angles. It may be true that different measuring techniques work in different situations.

The conclusion for the experiment in July 2005 is that string or computer reconstruction based on traditional concepts may apply but should not be used alone to draw essential conclusions regarding the origin of a gunshot. The results of computer applications, laser projections, or string reconstruction should be confirmed with other techniques until adequate comparison studies are conducted between gunshot distributed blood drops and bludgeoning distributed blood drops. Two techniques assist investigators in locating the general area of an origin: blockage (void) transfers and multiple surface areas of convergence.

Template-type transfers[8] show the location of an object at the moment of a shot, and may be used to help locate the position of an origin. To benefit from these, the objects need to be in place or recovered and repositioned. Lines, strings, or lasers may be used to project back toward the general area of the origin. Because ceiling beams, light fixtures, furniture, and drinking vessels may be present at crime scenes these can help indicate a general position of the origin. One must keep in mind that gunshots involve a V, cone, or triangular distribution, which may involve an apparent blockage that is in fact a true void or absence pattern rather than a blockage.

A good reconstruction of the origin includes multiple areas of convergence located on surfaces positioned at angles to each other, such as adjacent walls, ceiling, and floor. The convergence of multiple areas of convergence can suggest and confirm an origin located by computer, laser, or string techniques. The use of areas of convergences alone, however, locates a general position of the origin rather than

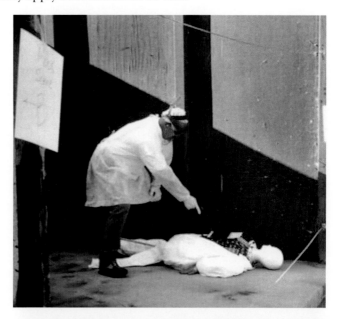

Figure 4-12

GDIS mock crime scene set up at CHP Academy near Sacramento, July 2005.

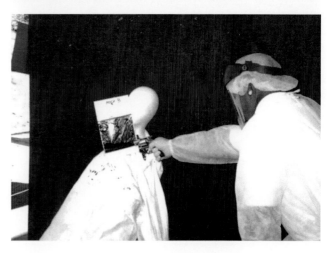

Figure 4-13

One of a variety of poses attempted in GDIS mock crime scenes.

---

[8]Wonder, *Blood Dynamics*, 85.

a specific locus of points, and should not replace the formal reconstruction. Still, confirmation is always beneficial.

With the excellent mathematical and physics minds working on this technique, the future will no doubt include updated principles and corrections of errors in reasoning. At the present time it is important to recognize that the technique does appear to work, as seen in hundreds of mock crime scene exercises. It is essential to not throw the baby out with the bath water. This can be achieved by using confirmations, areas of convergence and blockage patterns, and working on developing better mathematical approaches for reconstruction of the origin.

ANALYSIS, MOCK SCENE B, BPT METHOD (TIME AND CONDITIONS LIMITED TECHNIQUES)

25.7% error measured from the floor
2.7% error from east (longest dimension, behind shooter, placement of shooter)
25% error from north wall

ANALYSIS, MOCK SCENE D, OVAL

18.2% error from the floor
0.9% error from east wall (better area of convergence)
46% error from north wall (small value created appearance of greater variance)

ANALYSIS, MOCK SCENE E, BPT (BEATING)

Table 4-1

Results (in inches from) for July 15, 2005 Experiment at CHP Academy

	Mock B		Mock D		Mock E		Mock F	
	Actual	BPT	Actual	Oval	Actual	BPT	Actual	Oval
Floor	35	44	38 ½	31 ½	16 ½	34 ½	33	34
East Wall	73	71	111	112	105	Ok	95	89.5
North Wall	12	9	13	7	100	Ok	11	11.5

Participants didn't measure the location but views showed that the origin was located horizontally according to the area of convergence and blockage, which was correct. Location of the origin of the blood source vertically, however, was 18 inches too high. This represents an unacceptable deviation from the actual site of the impact between blood and hammer. In reviewing the reconstruction it was found that although participants could determine the exact location horizontally for the impact from the up-ended waste basket blockage pattern, projection of the angles to locate the origin included attempts to find a point source. Hitting the pool of blood distributed drops directly up as well as at angles. Some drops projected at angles were recorded on the wastebasket top that was removed and unavailable for students to use in the reconstruction. The stains they did use may have been drops directed up initially, which then fell at angles closer to 90 degrees after reaching their apogee. This suggests a need for caution in drawing conclusions from beatings where blows are delivered directly down, i.e., drops travel up, then fall directly down rather than distribute at true angles from the blood source.

Figure 4-14

Alternative pose to see if position of shooter could be recognized from distributed patterns.

ANALYSIS, MOCK SCENE F, OVAL

3.0% error measured from the floor
5.8% error measured from east wall
4.5% error measured from south wall

OBSERVATIONS OF EXPERIMENT AND RESULTS

One monitor used the class to teach. While this was not an objective of the experiment, it did present an opportunity to compare techniques, and was not discouraged. A common problem with string reconstruction was seen in different teaching techniques, drawing strings to a point. Although gunshot-distributed blood may lead back to nearly a point, all stains cannot originate from a single point. When an attempt is made to find a single position where all strings meet, the origin may be placed farther away than the true location. This was true of the gunshot scenes as well as the beating one.

Completing the oval is generally less desirable with more elliptical stains than BPT in measuring. However, with higher degrees, more open angles, completing the oval may be better. Comparison of measuring techniques as they would be used at actual crime scenes needs much more work. This cannot be

Figure 4-15

Untrained participants did a remarkable job on this scene.

Figure 4-16
Strings drawn to a point
extended origin too far
from actual location.

conducted satisfactorily as Drip Cast Offs on slanted targets. Hopefully instructors who have spent years compiling data from mock crime scenes will evaluate information as comparisons with future gunshot-distributed mock crime scenes.

RECOMMENDATIONS FOR REPEATING THIS EXPERIMENT

1. Do not do it outside on the hottest day of the year!
2. Fewer teams are better (e.g., 3 participants per team with a maximum of 4 teams, rather than 5 or 6 people per team with 6 teams).
3. Fewer scenes and more frequent trials will be better (e.g., 4 scenes repeated at 3 locations rather than 6 scenes at two locations).
4. It was good having food and drinks available on site so that the class didn't need to leave and return. Hand washing between the experiments and the lunch room need to be strictly enforced.
5. It is essential that the areas of convergence be done consistently. Having participants new to the evidence construct areas of convergence probably influenced the results in scene B. The monitor drew strings to a point rather than identify the size and shape of the convergence.
6. Having comparable equipment for all scenes is important.
7. Future trials should include premeasuring the selected stains to compare angle computations.
8. Individual packaged blood samples did speed up the setting of multiple scenes, but adding an emesis basin (available in hospital emergency rooms and clinical labs) under the pouch to catch drips immediately following the shot would help prevent use of information other than the spatters to locate the area of convergence). Quickly remove the blood source after each shot.
9. The use of recently donated but underfilled and rejected packed red blood cells (rather than whole blood) worked well. This would mimic a body gunshot to the liver, lungs, pancreas, and spleen areas. Head and heart shots might have a lower hematocrit but be harder to see and measure on dark walls.
10. The walls in our rooms were dark brown, which contributed to eliminating scenes from use. Coating to prevent blood sticking can be a problem also. White cardboard or white latex painted surfaces are always preferred in experimentation.

PROTECTING YOUR RESULTS FROM CHALLENGE WHILE NEW MATHEMATICAL FORMULAS ARE FOUND

POSSIBLE CONCLUSIONS AT THIS POINT

Figure 4-17

Contrasted BFIS with a hammer.

1. Quality of area of convergence was essential.
2. Due to variable error it is best to use confirmation techniques with reconstruction of the origin methods; i.e., observations such as blockage patterns and areas of convergence from multiple intersecting planes.
3. Location of the origin from the direction of the shooter is more reliable than other locations.
4. The least reliable location is from the floor for origins located on the vertical plane (this can be explained by point/scatter location problems).
5. At this point, with this very small sampling, it cannot be concluded whether the BPT or the oval method is better with firearms. More tests need to be conducted before a conclusion would be possible between these. For that reason it is probably best to keep both techniques in mind, rather than advocate the sole use of either method.

There is no doubt that the concept and mathematical principles are flawed and/or have been used out of context. Unfortunately (or fortunately) the technique seems to work. Although none of the scenes in the experiment were "right on," they are remarkably close to where the actual shot occurred. Deciding where the origin was positioned, however, was greatly influenced by observing things other than the measurements of the stains, such as:

- Areas of convergence
- Blockage patterns
- Blood drips and pools
- Transfers from persons and objects immediately following the shot
- Scenario story

All except the last item were helpful in checking and confirming where an origin actually was located. Because of the theoretical problems with the trig applications it is wise to confirm any results with other information possible with bloodstain pattern evidence. If challenges are made to the science applications of math alone, other points can save the testimony and confirm the origin

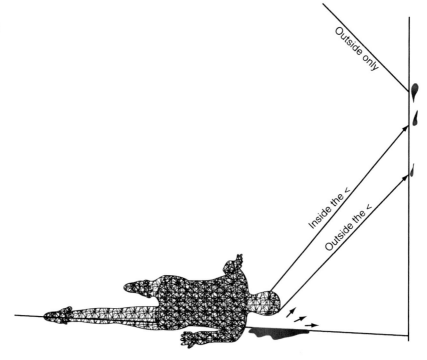

Figure 4-18

A stain traveling up a wall could result from an event inside or outside the angle, but a stain traveling down would have to originate outside the angle.

as concluded. This should be considered even with computer programs that locate the origin—or perhaps especially with those since the area of convergence may not be adequately appreciated as it is with a manual applied string (or laser) reconstruction.

MORE SUGGESTIONS IN RECONSTRUCTING THE ORIGIN OF AN IMPACT

1. Construct a good area of convergence: this cannot be overemphasized.
   a. Do not draw to a single point.
   b. Leave area of convergence lines in place when constructing the lines to the origin. These lines help align the base position of the protractor.
   c. Keep the line from the stain to the origin straight. Avoid holding the string to the angle.
2. Select the right stains to measurement:
   a. Visible with the unaided eye. Don't include something you can't see without magnification for an area of convergence, nor should they be used for angle determinations.
   b. Oval with distortion end, tear drop, tadpole, etc. Stain must have direction of travel consistent with area of convergence.
   c. Consistent with other stains selected to measure.
   d. All on one type of surface.
   e. Direction of travel is in the direction of the greatest edge distortion (not necessarily the longest measurement). Do not measure if the distorted edge is not the longest edge.

3. Measure correctly:
   a. Completing the oval is not the best choice (although it might be acceptable with more oval-shaped stains).
   b. BPT technique is better with more elliptical stains.
   c. Training programs with mock crime scenes are best for practice in measuring.
4. Do not draw either the area of convergence or the origin to a point.
5. Confirm with other information:
   a. Blockage patterns.
   b. Areas of convergence on two or more surfaces at right angles.
   c. Drip and cessation cast offs, transfers, etc.

INTERPRETATION

The traditional way of interpreting the results of a location of the origin is worded as "the event occurred below the *Point* [sic] *of Origin.*" This is applicable if the impact spatters used in the reconstruction are on a horizontal ground or floor surface and complete the oval is the method of measuring. The incident angle will be more open than actually occurred but the error will compensate for limiting the event height.

A problem occurs when the impact spatters used are not on a ground horizontal plane. In all other arrangements of the spatters used in an origin reconstruction consideration must be given as to whether the blood drop was going up or down when it was recorded as a stain. If the drop was going down, the wording would be that the event happened "outside" rather than inside the angle (Figure 4-18).

A common misconception is that blood drops distributed from an impact begin in arc flight paths (Figure 4-19). Blood drops from an impact begin in straight flight paths and arc after wind shear and gravity overcome forward momentum imparted by the force to the blood source. Untrained or incompletely trained individuals may assume the arc flight path explains contrary directions of travel, which in fact are more likely to be cast offs not impact spatters.

Figure 4-19

A common misconception regarding arc flight paths is that they begin the arc at the origin. This is not reality. The projection starts straight and bends as wind shear and gravity act upon the drop. The smaller the drop the less gravitational affect, but the less momentum to continue against wind shear. Within an enclosed room, blood drops distributed by gunshot probably do not bend significantly.

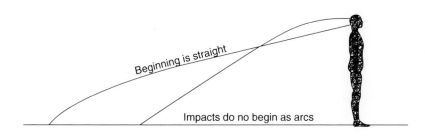

Beginning is straight

Impacts do no begin as arcs

NEW APPROACH TO ORIGIN RECONSTRUCTION

The worst technique for reconstruction of the origin is in drawing strings, lasers, or computer lines to a point, as a point of origin. There are at least four reasons to discourage this technique:

1. No two drops of an incompressible fluid, which blood is, can originate from the same point.
2. Providing a point suggests a greater sense of exactness than exists. This can lead to successful challenges in science as well as law.
3. Drawing lines, strings, or lasers to a point frequently projects the origin beyond where it actually lies. This too can be challenged in science and law.
4. Because impacts are never actually points and the size and shape of the origin reflects the size and shape of the actual impact, drawing lines, strings, or lasers to a point is a loss of investigative leads information. The actual size and shape of the locus of points which contains the origins of impact spatters can suggest the size and shape of a weapon and/or the sequence of separated impacts during an assault. It is essential in investigations that valuable evidence not be wasted.

Figure 4-20 shows a schematic outline to the common string reconstruction of the origin. Many improvements have been added to this technique. One point from experience, however, is that methods that do not use areas of convergence omit a valuable step in the procedure.

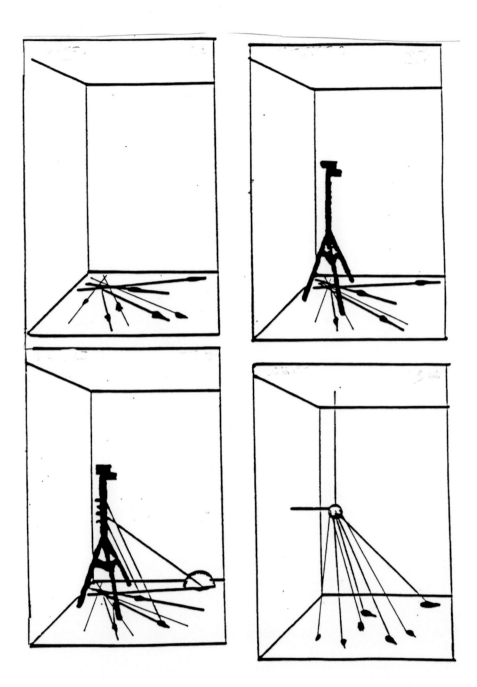

Figure 4-20

A schematic diagram shows the process of reconstructing the origin with strings.

CASE APPLICATIONS

The following cases were processed from photographs, occasional visits to the scenes (seldom fresh), and usually confirmed with autopsy, medical, and interview reports. The cases were selected solely for the purpose of academic review and the hope that my experiences will benefit future applications of bloodstain pattern evidence (BPE). It is the nature of forensic science that differences of opinion and interpretation will always exist. Two quotes come to mind with regard to an anticipation of criticism.

> If you come across error, rather than uprooting it or knocking it down, see if you can trim it patiently, allowing the light to shine upon the nucleus of goodness and truth that usually is not missing even in erroneous opinions. Pope John Paul I (1912–1978)[1]

> Nothing would be done at all if one waited until one could do it so well that no one could find fault with it. John Henry Newman (1801–1890)[2]

With the second quote it is accepted that others may provide different and perhaps even better conclusions. Such is the source of knowledge and the advancement of science.

[1]Found in *In God's Name*, by David Yallop. (1984). Bantam, 20.

[2]Found online in *The Ultimate Success Quotations Library*. (1997).

Figure 5-1   Chicken House IV, *by American Painter Eva Martino, 2004.*

CHAPTER 5

TIMING IS EVERYTHING

INTRODUCTION

Although the usual procedure is to look at the scene first, or photographs of it, the format for these cases begin with the identification of blood sources available. In practice, medical records and autopsy reports are not available until after bloodstain patterns have been classified into major categories and the preliminary events sequenced. In the academic use of material, however, it is convenient to know what blood sources were available. All cases should be approached after becoming familiar with the investigative tools in the appendixes.

There are many good procedure resources for approaching crime scenes in general. This work is to provide additional information regarding one form of evidence rather than restate what other well-qualified authors have covered. Each case is presented in a format that should make it unimportant whether the case was analyzed from the defense or the prosecution viewpoint. In fact, casework for both sides of the aisle has been processed by the author. Two cases were reviewed from one perspective and later reviewed for the opposing council. Mistakes were made at all levels—including by the author—but brilliant sleuthing also was found at all levels. It is the nature of humans, however, that we learn best from our mistakes and not from patting ourselves on the back. Nothing in this book is directed at the discomfort of any individuals encountered during my 30 years' experience with the amalgam called the Criminal Justice System.

FURTHER DISCUSSION OF VELOCITY IMPACT SPATTER TERMINOLOGY-SUBJECTIVITY

After Dr. Kirk's death from cancer in 1970, changes to the original semantics occurred. Instead of the impact and relative velocity being considered at the target surface, the meaning was shifted to the contact point between a weapon and a blood source/injury. This has caused confusion in part because the original concept of contact between a single blood drop and a target also was retained. So impact site became both Dr. Kirk's definition as the site where an individual stain was recorded on a surface, as well as the new definition of the area where a weapon opened a blood source to distribute drops.

Pretests, submitted to classes of students with little background up to completion of two or more 40-hour Bloodstain Pattern workshops, show that the velocity impact spatter (VIS) terms often are regarded in a subjective context. Although this has been corrected in most training formats, many participants still feel that VIS means a specific size of bloodstain identifies specific events, i.e., gunshot is identified by a specific size spatter (less than 1 millimeter in diameter) called high velocity impact spatter (HVIS). Beating bloodspatters are identified by specific size (1–4 mm) bloodstains called medium velocity impact spatter (MVIS). Impact events, however, involve a variety of drop sizes within each degree of force, and in fact are characterized by the presence of an array of sizes, never limited to a single one nor narrow range of sizes. Different pattern dynamics, impact, **cast offs**, and arterial damage, also distribute drop arrays with considerable overlap in stain size ranges.[3] There is no such thing as one identifying bloodstain size, nor narrow limit range, for an entire dynamic category. Patterns consisting of many spots of blood can result from different acts or events, not all are criminal.

Cast offs: Blood drops which separate from the surface of a carrier.

The main problem with using VIS terms as originating at an injury rather than as Dr. Kirk's approach of a blood drop at contact with a surface, is that we are no longer dealing with identifying individual bloodstains at a single location in time and space. By current methodology labeling is based on the collection of stains on a surface separated from the defining velocity event. This is a shift in application of the term in an analysis, from a single item of physical evidence at one point in time and space to the behavior of a group of spots between a highly variable event (not an item of evidence) and recording upon a distant surface with its own set of variable conditions as well as conditions between the two. To perform this analysis, the analyst must first assume a link between the two locations, contact with a blood source and recorded spatter pattern. Experience has shown that the recorded spatter grouping is not always from the assumed dynamic event. Shifting from a velocity component at the contact between target and blood drop to the contact of a weapon at a blood source, then reading the results of an arrangement of spatters on a target adds many variables that must be considered before conclusions can be stated with regard to the analyzed spatter. This is shown in Table 5-1.

Bloodstain patterns identified on the basis of velocity of a weapon striking a blood source involves so many uncontrollable variables that representatives of the scientific community now doubt that BPE can be considered a science. Many claim it is just police work.

More important to the future of the evidence is the shift in logic regarding recorded bloodspatter patterns. Dr. Kirk's logic, as understood from his lectures

---

[3]Laber, Terry. 1986. Diameter of a bloodstain as a function of origin, distance fallen, and volume of the drop. *IABPA News* 12:1, 12–16.

VIS per INDIVIDUAL Spatter	VIS per EVENT to a Blood Source
Blood drop size	Blood drop size
Velocity drop is traveling at contact	Velocity of weapon at contact
Target surface characteristics	Characteristics of the weapon
	Characteristics of blood source (hardness, presence of hair, clothing, fat, bone, skin)
	Nature of blood vessels injured
	Amount of blood distributed
	Degree of blood source break up
	Distance traveled to target
	Conditions between injury and target (wind currents, heat, obstructions)
	Velocity of array of drops at contact with target
	Target surface characteristics
	Overlap of other events
Only three variables are considered relative to interpretation of individual stain appearance	12+ different variables need to be considered before conclusions can be stated regarding the whole group of spatters

Table 5-1

Variables for Identification of Velocity Impact Spatter (VIS) Events

to the California Trial Lawyers, was to look at how the blood spots (spatters) could be interpreted to identify the type of dynamics that distributed them. He felt a way could be found to determine from their arrangement what act was involved, whether impact, cast off, or arterial. Dr. Kirk analyzed the bloodstains first, from which he felt someday one could postulate the type and condition of dynamics that distributed the whole array. The appearance of individual bloodstains could indicate velocity as one of many criteria. The bloodstain patterns of a case were treated as physical evidence, not as a conclusion to other investigative information.

When shifting the source for determining velocity, the logic changed to assumptions which weaken the evidence. Velocity as the key to identification of whole groups of spatters requires that the source be assumed before the pattern can be labeled. The revised approach deleted evidence from the initial consideration, and thus, became a subjective approach. The assumed dynamics on occasion also has ignored other noncriminal events, such as blood dripping into blood and respiratory distribution.

An apparent attempt to correct the lack of physical evidence in the identification is to claim that size of the drops from the break up of a blood source can provide identity of velocity. Unfortunately size is one of the variables that became considerably more complex when the site of velocity estimation was shifted. Bloodstain size is dependent upon many factors, and cannot be estimated from any assumed full-sized drop, so that it is scientifically unsound to claim size alone identifies an entire category of dynamic events. The size of a bloodspatter, also, provides information regarding that spatter, not the whole arrangement. The sizes of the associated stains within a group provide information regarding the whole pattern group if (and only if) it is first established that the group was distributed by a single event, specifically an impact. To do this one should first justify the identification of impact, not use the stain sizes to justify identity of the event and then use the event to identify the stains, a loop of unsupported reasoning. Crime scenes involve considerable overlap of spatter arrays. It must be shown that a group of spatters are not overlapping separate events, especially as often occurs at violent crime scenes that may include cast offs, arterial damage, respiratory projection, and explosive blood into blood as well as impact.

In consequence, identification applications may now be subjective, such as when an "expert" requires complete interview background before they can identify a pattern as MVIS or HVIS. With this type of analysis, a crime involving a gunshot would have any spatter pattern labeled as high velocity impact spatter. If no gunshot is stated, the identification will be medium velocity impact spatter. The result is the loss of investigative leads information derived from a complete bloodstain pattern analysis. The identification is based on hearsay, not upon independent criteria of the stains as evidence themselves. If a gunshot did not occur where speculated, or did occur but was unknown when the scene was processed, the interpretation of the bloodstain patterns may be discredited along with any erroneous earlier assumed scenario. In fact, a misidentification of a single VIS pattern can prejudice all other pattern identifications within a case.

It will always be more professional, and consistent with Paul Kirk's initial approach, first to label an arrangement as bloodspatters (blood spots, bloodstains), then identify major classifications as cast off, impact, arterial, respiration (**exhalation**, expiration), or **blood into blood**. Follow this with final classification in terms of velocity, if necessary, rather than leap to an immediate specific VIS term at first sight of a pattern; i.e., "That's medium velocity impact spatter!" It might be, but it also might lead to embarrassment in admitting later that the pattern was really **cessation cast offs**, blood into blood, respiratory wheeze, etc.

Exhalation: Blood drop distribution from respiratory functions like breathing and wheezing.

Blood into blood: Scattered secondary spatters distributed when blood drips into a volume of blood.

Cessation cast offs: A cast off pattern where blood drops are distributed at the moment the carrier stops.

CASE 1

BACKGROUND

A suspicious death was originally treated as a traffic accident investigation (TAI). The victim, a woman, apparently was killed as the result of her car leaving the

roadway and traveling down a short embankment into a dry creek bed (Figure 5-2). A passenger in the same vehicle (hired man of the
woman's husband) sustained minor injuries.
The vehicle following the victim's car was driven
by a rural volunteer fireman, trained as an emergency paramedic. He claimed that he was at the
woman's side within five minutes of seeing her
go off the road. Because officers questioned
the death resulting from the circumstances, a
more thorough work-up was requested on what
remained of the evidence (i.e., the car and
photographs). Photographs subsequently were
taken by an evidence technician experienced
with homicide investigations.

Since the location of this case was a small,
upscale rural community, homicide investigations were rare. The anatomical autopsy
was performed by a clinical pathologist at a
community hospital with the cause of death listed as a broken neck and manner
of death left as unresolved. As evidence in an open case, the car was stored in
the local government vehicle impound yard. This was an unpaved lot outside,
exposed to the elements, with only a loose canvas cover over the car.

Three years later, the hired man was in a bar, 3000 miles from the scene.
Both the victim and her newlywed husband had lived in the area before moving
across the country. The handyman bragged about helping dispose of his former
employer's wife's body. Relatives of the victim were in the bar at the time, overheard him, and subsequently contacted law enforcement officers.

Evidence from the original investigation was limited to reports, photographs,
and the car. The victim had been cremated, upon request of the husband. Clothing had been removed and discarded at the hospital upon admission, and the
body cleaned prior to the autopsy, where no photographs were taken.

Figure 5-2
*Slope down to a creek bed
where the victim's car
traveled.*

BLOOD SOURCES IDENTIFIED

There are basically three types of sources for blood at the scenes of violent crime:
injuries, volume blood (pools), and prior wet stains. Prior wet stains can be any
bloodstain that remains wet for sufficient time to provide a blood source for
subsequent staining. Usually these are limited to large drips, blood flows, or very
bloody transfers. During the time required to dry, transfer (contact, compression
stains) or **drip cast offs** (LVIS, drip trails, passive, gravitational stains) may occur.

Drip cast offs: Blood drops
distributed from the
surface of a slow moving
carrier.

Bloodspatters generally dry too rapidly to provide a good blood source unless someone steps into or rubs the array immediately following drop distribution.

The only external injury blood source noted in this case was a zigzag-shaped abrasion located between the victim's eyebrows, which was specifically noted in the autopsy report. It was also noted that there was no blood inside the mouth or any of the respiratory organs such as the trachea, bronchi, nasal passages, or throat. No exterior examination was provided regarding the body at autopsy, with the exception of noting the abrasion.

INTERPRETED FROM PHOTOGRAPHS

Figure 5-3

View of victim as found. Note position of nose upward from area between her eyebrows.

The predominance of other possible blood sources involved flows and transfers (Figure 5-3). The source for flows was limited to the abrasion between the victim's eyebrows. Blood was seen on her lips and left cheek, with some random spatters on her bra. Note that the photographs were taken after paramedics attached cardiac leads. Any bloodspatters seen would need verification that resuscitation efforts were not responsible for distribution of drops. No impact sites, cuts, or arterial damage areas were noted in the autopsy report or suggested from review of the photographs. The hired man (the passenger) had no broken skin injuries.

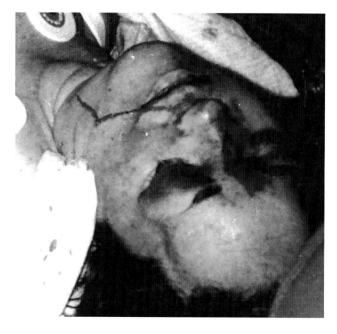

Wipe: A moving transfer pattern where blood was moved from the original appearance.

BLOODSTAIN PATTERNS IDENTIFIED

Moving and direct transfer (compression, **wipe**) patterns were found on the victim's face, in addition to at least three flow patterns, two of which had apparently been wiped. One flow pattern appeared to travel uphill from the abrasion to the right of the nose bridge, then wrapped around the nose tip to the lips where blood pooled to form a volume bloodstain. From the lips blood overflowed and ran down the chin in a direction contrary to the position the body was found. The victim was in the break pedal well with her head resting on the driver's seat. The flow would have been normal if the head were positioned up with chin down. The edges of the stain on the victim's lips reproduced the zigzag of the blood source, the abrasion, between her eyes. This suggested some object applied to the victim's face with a reproducible zigzag configuration.

Unclassified spatters (spots, stains) could be seen on the victim's bra, which were exposed by paramedics attempting to resuscitate her (see Figure 5-4). Emergency people noted the stains were dry on the bra when uncovered, before resuscitation efforts. They did not correspond to other stains seen on the blouse by emergency workers.

A dark colored smear of blood was noted under the ignition switch on the car during the first examination. Review of the driver's seat showed a few old black smears, which would be expected to be consistent with bloodstains without preservative after three years. From the photographs, shaded areas on the deceased's blue jeans were misunderstood originally as appearing to be a urine stain. It is probable that a urine stain occurred but the shadows seen on the jeans more likely resulted from the positioning of the light sources for photography. The pattern resembles a shadow from the victim's hand and the seat (Figure 5-5).

On a second visit to the car to verify patterns noted earlier, an arrangement of **swing cast offs** (cast offs) was found on the glove compartment cover (Figure 5-6). These were a bright cherry red color. A single drip of blood was also seen on the passenger seat (Figure 5-7, see page 57). After three years of weathering it is highly unlikely that blood would retain a bright red color without preservative. The drop on the passenger seat was also discounted as part of the original crime since the passenger allegedly was seated there during the time blood would have been distributed consistent with the accident. A request was made to officers to test the red stains to verify they were blood, were human blood, and were the victim's blood type or DNA.

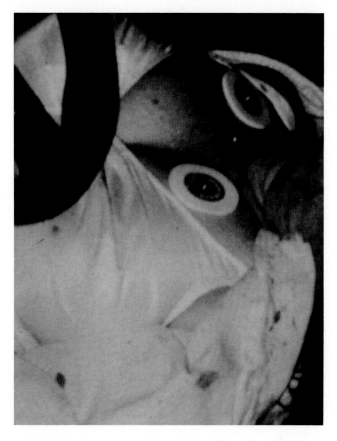

Figure 5-4

Spots seen on the bra do not match those on the folded back blouse. The volunteer paramedic verified that all bloodstains were completely dry when he reached the victim.

Swing cast offs: A pattern formed when a carrier swings in an arc in proximity to a recording surface.

OBJECTIVE APPROACH TO INTERPRETATION

Although the paramedics attempted to resuscitate the victim with vigor, no smears, smudges, or transfers occurred with blood that could have been distributed during these acts. The paramedics verified the bloodstains present were dry at the time they worked on the victim. The car had tilted upon impact

Figure 5-5

Shadow on the victim's jeans was called a urine stain, but note placement of light and appearance of enlarged view of the victim's hand.

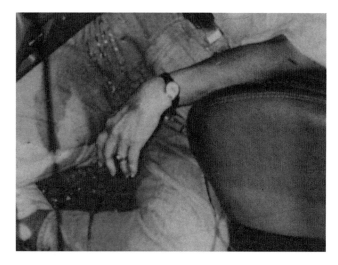

Figure 5-6

Cast off arrangement on glove box cover was out of context with other stains in the car as well as out of context with the alleged accident.

with a small tree before coming to rest. The ground temperature was cool, but not cold, with some unrecorded humidity. The stains seen on the victim's bra resembled random drip cast offs (LVIS, gravitational, passive staining) more than a distributed spatter from impact. The flows and wipes showed initial directions of travel consistent with the victim upright with head bent forward and blood flowing from between her eyebrows, along her nose, and accumulating (pooling) in her lips. None of the flows were consistent with her final position on the floor of the driver's side of the car as found immediately after going off the road. In the final position as the body was found, the flow on her nose appears to be uphill before trailing to the lips.

The shape of the volume stain (pool) on the lips matches the zigzag patterns noted elsewhere. There were three separate zigzag patterns that appeared to be a sequence of transfers (contact, compression), where the same shape transferred blood to separate areas on the victim's face. The article suggested for the pattern was a watch band, possibly used as brass knuckles over the fingers to hit the victim. The judge requested specific reconstruction showing the pattern, which was not available. Testimony regarding experience with such items in the presentation of 40-hour workshops was permitted.

An expansion band watch was collected when the husband was arrested, but was released to his family soon afterward. It could not be located later. No other article with the zigzag pattern could be identified within the car.

PUTTING CASE MATERIAL TOGETHER

Areas of concern in preparing the material for trial, and thus a condition of the report included:

- Was the material in the photographs blood?
- What would be reasonable drying times for flows?
- Were the bloodstain patterns consistent with a traffic accident?

WAS THE MATERIAL IN THE PHOTOGRAPHS BLOOD?

Bloodstain pattern analysts frequently are asked to verify if patterns seen in photographs are indeed bloodstains. It is argued that unless chemical methods are applied, one cannot tell if it is blood; and if it is blood, whether it is human or animal; and if human, if it is the victim's blood. The primary guideline for interpreting blood from photographs alone is that there will always be situations where you cannot tell if the substance is blood from a photograph. There are, however, techniques that improve confidence levels in regard to concluding stains as blood when the actual substance is no long available.

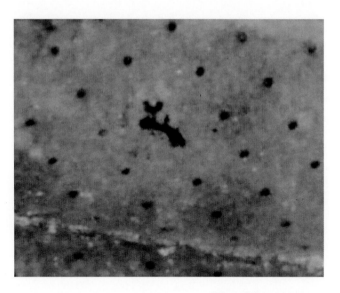

Figure 5-7

One drip was on the seat where the passenger sat. The color is wrong for blood years later, but also is discounted because the passenger sat there. It is suspicious from lack of smudging, direction of travel, and other drips or smudges around it.

ABC Verification of Blood Substance from Photographs[4]

The first requirement, of course, is that quality photographs be taken. Black-and-white prints have been used but are far less desirable than good quality color, glossy prints. Matte finish reduces blemishes and is preferred for portrait photography but should never be relied upon when specifically looking for spatters. If spatters are an issue and matte finish has been used, it is advisable to request glossy prints from the original negatives before basing testimony on the prints. There is debate regarding the use of digital versus film. The author prefers to leave such discussions to qualified crime scene photographers.

Appearance, behavior, and context may be used to improve confidence regarding whether stains seen in a photograph are blood. Appearance refers to color, hue, and tint of a portion of the photograph. To demonstrate how this can aid a person in identifying blood an exhibit was constructed using blood and substances which were alleged as possible sources for the red stains in the photographs (i.e., red-pigmented cough syrups; Figure 5-8). Although the exhibit was not required at trial, it provides an example of appearance verification. If a stain is identified by a person experienced with seeing bloodstains (medical personnel, paramedics, homicide detectives, evidence technician) with regard to color, tint, and hue, it is probably blood.

Behavior refers to blood's particulate composition and the ability to separate, clot, **hemolyze**, and behave as a result of the nature of its composition.

Hemolyze: The rupture of red blood cells to spill red pigment into a liquid of lower specific gravity.

---

[4]Wonder, *Blood Dynamics*, 11–12.

Recognition of such behavior, which may include characteristics of appearance upon drying, allows the examiner to separate blood substance from many other fluids of similar color.

Arterial spurt: An arterial pattern that resulted after a column separated into drops.

Context refers to identification of parts of a pattern. If a few spatters have been tested for blood and they are part of a complete impact spatter (spatter, stains), cast off (cast offs and passive stains), **arterial spurt**, etc., series of individual stains, then other stains from the same whole pattern group will also be blood.

The author does not use bloodstain-enhancing techniques. The general guideline is that if the bloodstains are not visible, attempts to make them so may alter them, or the viewer's recognition of them, to the point that errors in identification may occur. If occult blood testing or enhancement is to be done, it is left to law enforcement agencies to use their own protocols.

A few of the stains seen on the victim's blouse were identified by the paramedics as blood. Within the context, the other stains on the victim's underwear were also blood, of the same appearance, and the spread through cloth of the same behavior as blood. The stains on the bra did not appear to have corresponding stains on the underside of the blouse. This suggests that the victim was dressed after the blood source was opened.

Figure 5-8

Is it blood or cough syrup?

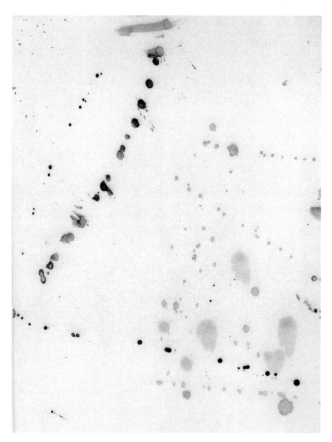

WHAT WOULD BE A REASONABLE DRYING TIME FOR FLOWS?

Because the drying time of the flows may have been viewed as important, an exhibit was made to illustrate timing for flows to leave faint traces after being wiped. A reasonable facsimile was approximated for the humidity and temperature of the stream area where the vehicle travel terminated. The stream area was cooler and probably a little more humid than the test area for the flows. The conclusion would therefore be that drying would be within that time or longer. Exact times for the experiment are purposely not given. When a court exhibit is necessary to demonstrate drying times, the experimentation should be conducted using the parameters of the specific case. In this situation the result of the exercise was that five minutes was insufficient for the degree of drying noted in the photographs of the victim's face. The exhibit confirmed that the blood flows occurred, were wiped, and dried over a time period in excess of that available between the alleged traffic accident and the arrival of the volunteer fireman. No transfer material to wipe the flows was found in the car.

Conditions of the reconstruction were that fresh human blood, hematocrit 47 percent, was drawn by syringe venipuncture without anticoagulant. Runs were immediately applied to the reverse side of pigskin suede, which is a reasonable facsimile to human skin, tilted at 35 degrees, at a temperature of 68° F. At one, two, three, four, and five minutes, respectively, a latex gloved finger was passed over the stain. At remaining times a facial tissue was rubbed through the stain. For light touching, five minutes was sufficient drying time but for wiping of the stains, up to 20 minutes was required under the conditions of this experiment. After six and a half minutes the whole sheet was tilted at 90° to see if the flows could change directions that late (Figure 5-9).

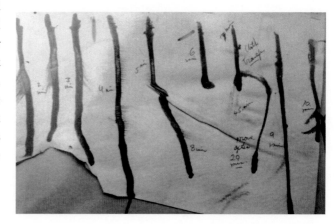

WERE THE BLOODSTAIN PATTERNS CONSISTENT WITH A TRAFFIC ACCIDENT?

The totality of the information from the bloodstain pattern evidence, traffic accident report, the witness volunteer fireman, and medical injuries report did not support the death as occurring at the time of the traffic accident. The conclusion from all evidence at trial was that the accident was staged to hide a prior homicide.

ADJUDICATION AND RESOLUTION

The hired man was convicted of second degree murder, and the husband was convicted of first degree murder in a separate trial. After a lengthy appellate process judicial errors were found in the second trial. The hired man at that point was out on parole and stated that he had done the killing all by himself. The District Attorney's office declined to retry the husband for the murder of his wife. Crimes committed at the jail while awaiting sentencing were added to his time served, keeping him incarcerated longer.

Figure 5-9

The smooth, pore side of pig skin suede was used with freshly drawn human blood, no anticoagulant. Blood was allowed to flow down the hide, then wiped at various intervals. The time to dry might be shortened with less humidity or warmer temperatures than were at the scene. This means the conclusions could be stated as "at least" the time period used was necessary for the effects seen in the experiment.

An interesting sidebar to the second trial centered on the red stains noted during the second visit to the vehicle. The swing cast offs (cast offs) seen on the glove compartment lid were out of context with both the homicide and the accident. No mention of the stains was listed in the bloodstain pattern report or brought out in direct testimony. The defense immediately began cross examination regarding those stains. The judge excused the jury and then allowed questioning to continue. When it was stated that those stains were ignored because the request was to analyze patterns related to the death only, the jury was brought back.

The entire car, interior and exterior, had been taken apart and each piece photographed before prolonged storage. The stains found on the glove compartment and passenger seat were not in the original photographs, thus confirmed as not part of the alleged crime. Had erroneous interpretation occurred with the stains related to the victim or passenger, it's possible the entire bloodstain pattern report could have been excluded at trial. The bloodstains were credited as essential by the juries in both men's findings.

After being alerted by the bloodstain pattern analyst, the detectives sent the stains to the crime lab, which identified the blood as cat blood preserved with EDTA (ethylene-diamine tetra-acetic acid). This explained the bright red color as this preservative is both a blood anticoagulant and color preservative. The detectives were encouraged to believe that a feral cat had cut itself entering the car and flicked its injury at the glove box. Such cat behavior is not likely, and confirmation bloodstain patterns (transfers, smudges, drips) of an injured cat in the car were not found. Although EDTA can be an ingredient in many vegetable food products, it is not an ingredient in pet foods. In this case the source was never identified.

Because the jury was present at the beginning of the cross examination, the judge ordered an explanation of the defense attorney's question regarding the "other stains" found in the car. The witness was allowed to explain that the stains were cat blood, which were not mentioned in the report, but were in later photographs in evidence and thus available to the defense for review. The defense, however, was not permitted to cross-examine regarding the cat blood stains since they were not brought out on direct examination. The prosecutor then entered into evidence the individual original photographs of the car showing no stains on the glove compartment after the crime but apparently distributed at a later time. The defense was able to address the issue through the photographs. He had no further questions.

LESSONS LEARNED FROM THIS CASE

Sadly, some investigators advocate looking for evidence of a beating or gunshot and decide to omit bloodstain pattern evidence altogether when none are found. A number of things can be gleaned from this case. One is that traffic accidents can be staged homicides, and that good photographs may be as essential for TAI (traffic accident investigation) as they are with obvious homicides. Had the excellent photographs not been taken of the initial "accident," this case might not have been resolved.

With regard to the adding of cat-blood-distributed swing cast offs on the glove compartment, it is unwise to try to fit every pattern observed into a preconceived scenario. The objective approach starts without a scenario, which is why it is best done initially at the scene as early as possible, before a scenario becomes a focus of the investigation:

1. Describe bloodstain patterns (for later justification of identification).
2. Identify patterns on the basis of criteria (refer to the appendix).
3. Locate blood sources available (this was crucial with the cat blood patterns).
4. Identify dynamic acts that could have distributed the patterns found.
5. Sequence events.
6. Eliminate those events that are not possible, or contraindicated with other physical evidence.
7. Do not label bloodstain patterns on the basis of other evidence (especially witness statements).

The cat blood stains were out of context with the other patterns. Verifying that this was blood when it turned out to be animal blood in a preservative helped protect information contained within the bloodstain pattern analysis.

Figure 6-1   Wall, door, and wardrobe that were the focus of bloodstain patterns in the assault on Pamela MacLeod-Lindsay.

CHAPTER 6

THE ALEXANDER LINDSAY SECOND INQUIRY[1]

Case 2

INITIAL CONTACT IN AN INTERNATIONAL INQUIRY

Following a presentation on Arterial Damage Bloodstain Patterns for the Forensic Science Forum in Queensland, Australia, 1988, a forensic pathologist approached the speaker with a request for review of an ongoing appellate case from New South Wales, Australia. No promises were made because the case mentioned apparently had been reviewed by some of the best-known experts in the field. A photographic view of the scene provided an array of patterns (Figure 6-1).

The book, entitled *An Ordinary Man*,[2] contained crime scene photographs and was provided with various photocopied transcript material from the original trial and a previous inquiry (Australian equivalent to the U.S. Supreme Court hearing). Upon a brief study of the photographs it appeared that some pertinent bloodstain patterns had been overlooked throughout the history of the case. The significance would not be clear until much later.

On the 14-hour nonstop flight back to the United States, all the material tendered was read. Immediately thereafter, a request was sent to the New South Wales Public Defenders Office for the rest of the documents mentioned in the material read. Over the next few months, stacks of Australian/British-sized legal photocopies were received and reviewed. The results of reading and rereading this material and examining photos yielded a 52-page report followed by a 12-page addendum with additional notes. The overwhelming conclusion was that Alexander Lindsay could not have committed the crime as stated by the police, nor should he have been found guilty as tried in New South Wales Criminal Court. Communication during the following year led to the scheduling of a second inquiry, which was the first time such had been granted in Australian jurisprudence.

Although the main reason for the second inquiry was worded by the granting judge as, "I am of the opinion that there was a reasonable opportunity for

---

[1]Copyright in the material is reserved to the Crown in the right of NSW. No part of the material may be reproduced by any process or in any form or otherwise communicated as to any other person without the prior written permission of the Crown, except as permitted under the Copyright Act 1968 (as amended).
[2]MacLeod-Lindsay, Alexander. (1984). *The Ordinary Man*. Hale & Iremonger, Australia.

Mr. Lindsay innocently to acquire bloodstains on his jacket,"[3] a claim of new evidence also was entered on behalf of bloodstain pattern analysis (bloodstain dynamics phrase used during the inquiry in honor of Paul L. Kirk).

CASE BACKGROUND

On September 14, 1964, Alexander McLeod-Lindsay went to work in a "hotel" (Australian for "restaurant") in a suburb of Sydney. Throughout the night, he worked between two to three areas: a restaurant, a public bar, and a private party. He had not known that he would be working on that night until after 8 P.M., the time he was asked to check in with the manager. It was verified that Mr. Lindsay took one 15-minute cigarette break, during which he was asked for a favor from two NSW police officers. They wanted him to bring a customer outside from the bar so they could arrest him away from his mates. Mr. Lindsay would have had another 15-minute break but no one saw him take it, and later in the turmoil of the time, he couldn't remember if he had or not.

At about 9:30 P.M., someone parked a car down the street then walked to the Lindsay's home and entered through an unlocked trapdoor. Mrs. Lindsay awoke and was moving toward the noise to investigate when she was attacked and savagely beaten with a jack bit. The 24-inch, 7¼ pound weapon (Figure 6-2) belonged to the Lindsays and normally was used to prop open a separated garage workshop door.

After midnight Mr. Lindsay left work and drove home. The porch light, which was normally left lit for his return, was out as were all the lights in the house. The fuse box was in the separated garage. Lindsay entered the kitchen and removed a flashlight from the top of the refrigerator, then proceeded down the hall toward the bedrooms (Figure 6-3). The flashlight beam shown on his wife, resting against a large wardrobe in the farthest bedroom but facing the length of the hall.

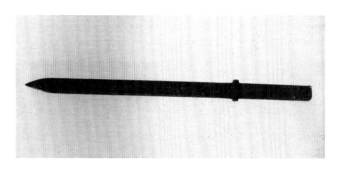

Figure 6-2

The weapon used to beat Pamela MacLeod-Lindsay was a bit from a pneumatic road drill. It was measured at 24 inches long and weighed 7 ¼ pounds.

From the extent of her injuries later described in medical records, she probably didn't look human at that point. Her nose was gone. Her cheeks were shattered, and brain pulp was oozing from her forehead. She was covered with so much blood early observers thought she was naked, when in fact she was still dressed in her pajama top.

Lindsay alleged that he had rushed to his wife's body and pulled her to him, exclaiming "What happened to you?" He laid her back and then ran to the neighbors to phone for help, since his own phone was out of order. When he returned, the

---

[3]*Lovejoy, The Honourable Mr. Justice. (1991). Report of the Inquiry held under section 475 of the Crimes Act 1900 into the conviction of Alexander Lindsay (formerly Alexander McLeod-Lindsay), 53.*

neighbors followed directly behind him and all later related what they saw. Lindsay stooped next to his wife and held her head and shoulders, then repositioned her on the floor while he crouched near. His 3-year-old son came to him for comfort. It was discovered that the son had been assaulted also, although apparently not as severely as the wife. Lindsay lifted his son, then walked with his wife on a stretcher to an ambulance. Afterward it was alleged that he reached into his car and removed a jacket after he had held his wife and son.

The police immediately focused on the husband as the guilty party and built what they felt was an adequate case against him. Unfortunately the standard police approach to bloodstain patterns at that time was a predominantly subjective approach. A conclusion was made of guilt. The evidence gathering was for the sole purpose of building a case against the husband rather than as an unbiased view for investigative leads. The police are not faulted for their actions in 1964, because this was accepted procedure at that time, and several points of the assault indicated knowledge of the house and family, which would fit the husband well. Mr. Lindsay's original statements were also inconsistent, later recognized as partial shock at finding his wife as he did.

Figure 6-3
The view down the hall from the front door and kitchen. Mrs. Lindsay was sitting on the floor propped against the wardrobe seen at the end of the hall. She probably didn't look human.

There was no scientific protocol to identify specific bloodstain patterns at the scene. Instead officers would "reconstruct the crime scene and look for pattern matches to crime scene photographs." The usual process was to use a blood soaked sponge, beat it with a weapon (in this case not even a similar weapon), and look at the results for spots of the same size as found at the crime (Figure 6-4). Directions of travel and arrangement of spatters within grouping of spots were not mentioned with regard to the scene, and most important, the fact that several patterns overlapped was not acknowledged (Figure 6-5, see page 67). Twenty-four years later a bloodstain pattern analyst with considerably more experience with blood but less crime scene and court experience was requested to redo the case.

BLOOD SOURCES IDENTIFICATION

An essential part of this case was that the medical records were excellent and available. Over many years and review by many highly qualified experts, no one

Figure 6-4

The police reconstruction shows "attacker" facing the wall hitting a blood sponge overhand. The attack was alleged overhand to the left.

Physiologically Altered Blood Stains (PABS): Evidence of physiological changes occurring before the stains were recorded.

PABS/mixed-CSF: A stain formed after blood mixed with another substance.

seems to have considered what happened to Mrs. Lindsay, instead focusing on describing how her husband assaulted her. All three general blood sources were available: bleeding injuries, still wet prior bloodstains, and volume accumulation (pooled) blood.

BLEEDING INJURIES

There was extensive damage to the face with superficial scratches to the head. Since the victim lived, any injury could continue to provide fresh blood over the time period between the assault and when she was placed into the ambulance. The possibility of clotting with reopening of the wound for more bleeding existed. This would contribute to bloodstains in and around the body, and could be distributed when Mr. Lindsay moved the body within sight of witnesses, when the paramedics worked to prepare the victim for transport, and/or while Mr. Lindsay accompanied his wife through admission in the hospital. It could also mean a crime of long duration with breaks in between assaults.

PRIOR WET BLOODSTAINS

Indications of both dried and wet stains were noticed on the bedroom wall and door, identifying events that occurred at different time periods. A fragment of clot can be seen sliding down the door from one of the elongated streaks (Figure 6-6, see page 68). This indicates a time span with different events overlapping. No notes were made or photographs taken of the area behind the door. Even if there were no stains there, it would help in sequencing and positioning the events that did leave bloodstains. It is best not to assume, when there is nothing there, that you don't need a photograph. Even with notes it is a good idea to have photographic corroboration.

At least three large areas of staining on the carpet (Figure 6-7, see page 68) showed variations in times and types of bleeding. One area is noticeably lighter than the others, which could be associated with loss of the cerebral spinal fluid (CSF) envelope mentioned in the medical reports. A clear fluid such as CSF, which was compatible with blood, would dilute it causing the lighter stain. This is classified as **PABS/mixed-CSF (physiologically altered bloodstains/mixed)**, in this case mixed with cerebral spinal fluid. One of the darker stains, behind the CSF/blood, shows considerable satellite staining around the circumference. This indicates rapid or forceful blood into blood (blood dripping into blood, passive staining, gravitational into blood) suggestive of **arterial gushing**. The medical reports also listed arterial injury. Bone fragments are noted in the photographs, and mentioned in reports, as associated with the volume stains. This could result from active bleeding cleaning the wounds from extrusion of driven- in bone debris when the victim was moved by Mr. Lindsay.

Arterial gushing:
A pattern projected from an arterial blood column which strikes a surface before separating into drops.

Although there were injuries to her shoulders and stomach, seen as bruises and tissue damage, blows struck there did not break the skin. Injuries not breaking the skin do not expose a source of blood. All of the blows responsible for distribution of blood were to the face and head. Lack of defensive injuries to her hands also suggested that she was unconscious from the first blows to the head, thus not defending herself.

BLOODSTAIN PATTERNS IDENTIFIED

Every classification of bloodstain pattern can be found in the evidence. Later, critics would dispute that the justification for the second inquiry was based on new evidence. It was an expressed opinion that the bloodstain pattern evidence (BPE) was not new at the time of the first trial. Although a good part of the pertinent evidence was established by Paul Kirk and technically available in 1964, the approach to identification and definition with subsequent sequencing to arrive at conclusions was a new, more scien-

tifically sound technique. The bloodstain patterns at the scene in fact had not been analyzed. A reconstruction was performed that varied substantially from the crime alleged, and the analysis had been left to the judge and jury in both the trial and first inquiry. Neither judge nor jury had sufficient background to interpret and draw conclusions regarding BPE. An objective approach to bloodstain pattern analysis, rather than the subjective application of **spatter analysis**, justified the claim of new evidence.

With a scene showing all the possible major classes of BSE, it is unwise to attempt to define every spot of blood. There were two predominant arrangements of spots that could be identified and labeled as to their possible dynamic source. The arrangements were both consistent with arterial projection (Figure 6-8). The first arterial pattern is the V arrangement of ovals seen in the middle of the

Figure 6-5

The actual crime scene shows considerable overlap of patterns expected from a beating such as this. Vicious beating crimes with arterial involvement are complex patterns.

Spatter analysis: An incomplete labeling of bloodstain pattern analysis.

Figure 6-6

A fragment of clot slides down the door from one of the big streak stains at the scene. This resembles the vertical cast offs of the reconstruction, but the position and dynamics were not a match for the actual crime events.

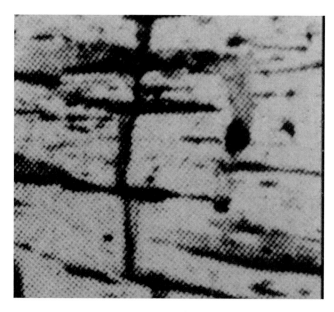

Figure 6-7

The different shades and view of the carpet pattern through the stains show that one was probably diluted with CSF. Note bone fragments, which could plug the injured artery. Clot would form around the bone and breach, to be dislodged when the victim was moved in front of witnesses.

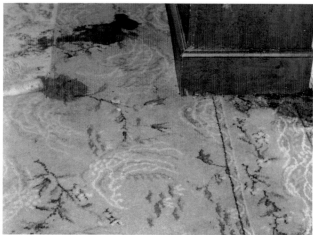

Figure 6-8

Arterial damage patterns. Large ovals in symmetrical V arrangement perpendicular to the floor. Streaks showing extension and elongation before separating into drops.

space between the door and wardrobe. This is identifiable as arterial damage on the basis of:

- Large equal-sized ovals
- Arranged symmetrically on each side of the V point
- Ovals perpendicular to the floor, not slanted as from a weapon striking a blow

To the right (as viewed) are streaks extending from the area near the V. It is noted that the type of material making up the jack bit, smooth iron with no rust, does not hold blood well. The conclusion follows that for the amount of blood within each arm of the V, the blood source is projecting relatively large aliquots of blood, almost as columns, rather than separated drops. The action that projects blood columns that subsequently may separate into drops is identified as arterial gush.

The second pattern group identifying arterial damage is the long streaks projecting from the V arrangement but located on the door. A fragment of clot flowing down the door is part of one of the streaks. This indicates that at the time of the arterial projection there was clot material within the injury, i.e., assault had to occur previous to the projection with a lag time sufficient for blood to clot to a firm stage *undisturbed*.[4] Although clot initiation may occur within less than a minute, detecting such requires the instrumentation of a clinical laboratory. For a firmed visible lump to form, the clotting stage would need to be greater than two minutes with no motion breaking up and disturbing the process. The medical indication of what happened is that bone was driven into the artery. The vessel constricted around the fragments and formed clot material. When the victim was moved, the bone was dislodged and arterial gushing distributed drops that became stains which included bone fragments.

Three of the eight experts consulted in this case identified blood clot material on Lindsay's jacket. One of the remaining five experts claimed the clot-appearing material was aspiration of blood along a fiber of the jacket. It should be noted that liquid may be aspirated. Red blood cells are particulate and are not usually aspirated along fibers, especially if clotting was also occurring. The clot stain flowing down the door confirms that clotting was occurring when the streaks, containing the clot, were projected at the door.

Another claim was in relationship to the size of spots on the slacks worn by Mr. Lindsay. Various actions will distribute small and fine-sized blood drops. These include blood dripping into blood, exhalation (wheezing, breathing, sneezing, expiration, respiration), and blood ricocheting (bouncing) off

---

[4]Wonder, *Blood Dynamics*, 104–109.

structures such as the wheels of the medical/hospital gurney or furniture at the scene. A volume stain (pool) existed at the location where Mr. Lindsay was seen crouching by his wife, cradling her head. This spot shows satellites of blood into blood (secondary spattering) around the circumference of the volume stain. Mrs. Lindsay was alive and still breathing when found and when moved to the stretcher to transport to the hospital. Damage to her face and nose confirms that she would be exhaling blood. Alex was seen by three witnesses crouching next to his wife next to the volume stain showing blood into blood characteristics.

OBJECTIVE APPROACH TO INTERPRETATION

Not observed in any original report but brought out at the second inquiry was the possible existence of a **transfer pattern** on the carpet. A pattern resembling the shape of the weapon handle was seen to the left (as viewed) of the stains in front of the wardrobe. The victim's pajama bottoms were thrown to the left of the assailant after being removed. There is a major bruise to the victim's right shoulder, which would be opposite the assailant's left hand in a face to face. The child had an injury to his head suggestive of a left-handed blow. Nicks in the wardrobe (Figure 6-9), suggest a left-handed overhand swing (Figure 6-10). The lack of blood in the grooves also suggested that the weapon was not bloodied on the tip when these nicks were made.

Although several indications existed that the attacker was left-handed, the police originally assumed right-handedness based, perhaps, on the husband

Transfer pattern: A pattern resulting from blood on one object transferring characteristics to another material.

Figure 6-9

Nicks in the wardrobe were overhand. Left-handed is suggested, but more important is lack of blood within the gouges. The weapon tip was not bloodied at that time.

Figure 6-10

How the nicks to the wardrobe could be delivered with left-handed overhand blows.

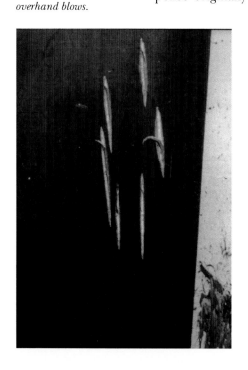

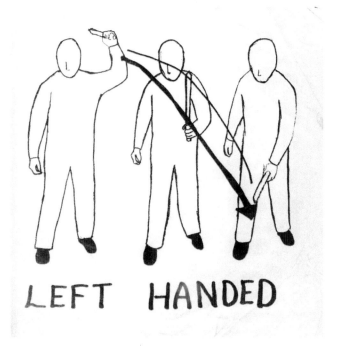

LEFT HANDED

being right-handed. Justification for the assumption was made regarding the blows being struck with the left requiring the right shoulder to be against the wall in order to deliver the blows (Figure 6-11). This does not follow as the only, nor even the most comfortable, position. The angle suggested is with the assailant facing the wall and swinging the weapon horizontally (Figure 6-12). The police reconstruction was manufactured in part with the attacker facing the wall, not as alleged, facing the victim, but delivering blows right overhand.

The most important part of the interpretation needs to address why Mrs. Lindsay survived such a violent beating. Even though the case was reviewed by many highly regarded experts in BPE, none appeared to take an interest in the excellent medical reports and care she received. The injuries were consistent with a horizontal attack, not a vertical one. Overhand blows of the magnitude involved should have killed her, whereas horizontal ones performed a frontal lobotomy. Mrs. Lindsay's medical follow-up over the many years following the attack has been consistent with this suggestion.

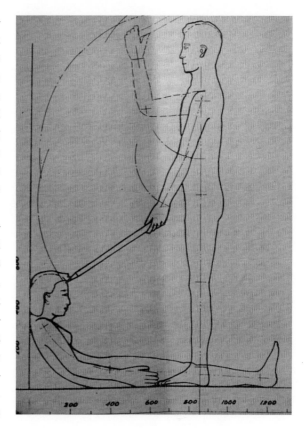

Figure 6-11
Computer model suggested by another expert in 1988.

Much was made of the clot stains on Mr. Lindsay's jacket, but a point that was not brought out until later was that for clotting to progress to extrusion of colorless serum, the clotting process must be left alone. One cannot have a beating occur concurrently while blood is clotting. It was difficult to convey that blood on the fabric of the jacket would not then be able to clot. The problem is that clotting requires a pool of blood. Such could not stay on a jacket being worn. Blood would flow off, especially on a jacket made to repel rain, as this one was. If the jacket was laying flat and a pool was on the surface, contact with fabric either prevents retraction or draws material into the clot rather than extruding serum in the normal way. The continued movement and disturbing of a clot will extend and prolong clotting. Had this delay in clotting happened, the serum stains would not have been found on Lindsay's jacket. Clot retraction and extrusion of serum had to occur before drops were formed and distributed on the jacket.

Some of the stains are so clear as to contain little or no red blood cells (Figure 6-13, see page 73). The shade of the jacket is visible through the stain without any shading. These also have dark lumps of material consistent with solidified clot fragments. For full retraction the clinical standards are one hour. With jostling and movement more than an hour would be required, or it could even be prevented completely.

It has been said that a blood drop falling into a pool is limited to bouncing back no more than one half the length of fall. This must be qualified regarding how a drop is projected toward the pool. Arterial gushing can project blood several feet out and above the level of the blood source. Blood into blood is governed more by the frequency of the drops, and/or force behind their dispersion, than by the height a single drop falls.[5] Arterial damage from vessels of normal blood pressure was confirmed with notes in Mrs. Lindsay's medical reports.

The primary evidence initially was claimed to be the jacket worn by Lindsay. Spots were found at various locations with **Direction of Travel** and the "size of match heads" noted. An observation was made that there was a portion where spots were not found. The inside bend of the elbow shows blockage of spatters as does the topside of the cuff. It would seem impossible to hold the heavy weapon and keep the elbow bent while beating someone. A beating blow would also be expected to involve blood deposited with strong directions of travel up the cuff. Directions of travel seen by the initial examiner was toward the victim not away from her. A press photograph caught Lindsay with a cigarette and his right arm bent in much the same way it must have been when he was exposed to blood drop distribution from the victim in order for the elbow blockage pattern to be recorded (Figure 6-14, see page 74).

Direction of Travel: The direction a drop of blood was traveling when recorded.

Later the importance was shifted to the slacks worn by Mr. Lindsay. These were originally analyzed purely on the basis of the size of spots seen. When a reporter later requested a view of the pants it was learned they had been discarded. They were not available for the second inquiry. The size alone of spots cannot be used to identify the dynamic act that distributed the blood drops. Too much overlap and variation occurred in defining the size of spatters. The most important observation regarding the pants, however, is that all the spots noted by the New South Wales examining officer were below the knee (Figure 6-15, see page 75). Only two spots were seen

Figure 6-12

A more realistic view of shift to horizontal blows when overhand hit the wardrobe instead of the victim.

---

[5]Wonder, *Blood Dynamics*, 134–135.

at knee level. Lindsay was seen by three witnesses crouching on his haunches next to the
blood volumes with satellite spatters (Figure
6-16).

The analysis contradicts the accusation
that bloodstain patterns are consistent with
Lindsay assaulting his wife. The total time
limit of 15 minutes was not sufficient for him
to remove his jacket from in front of the barkeep, drive home, park down the street, walk
to the house, enter, attack his wife, wait no
less than two minutes (and up to one hour),
repeat the attack, drive back to work, leave
his jacket under the bar without the barkeep
noticing, and return to work.

Considering all the ways that bloodstains
of the size ranges listed in this case could be
acquired, and pattern evidence confirming
that the conditions for these ways existed at the scene, it was an error to conclude that there was no other way to explain the bloodstains on Mr. Lindsay's
clothing. The bloodstain patterns are all consistent with Mr. Lindsay's original
statements to police.

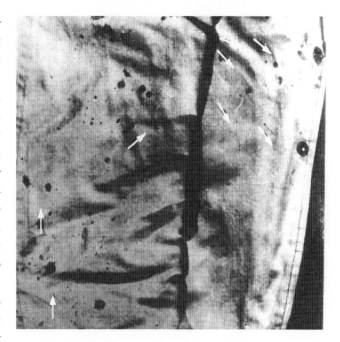

Figure 6-13
*The initial examiner
did note directions of
travel on the jacket but
appeared to not know
what to do with the
information. Early inter-
est in bloodstain pattern
analysis provided fasci-
nation with the evidence,
but sometimes how to
interpret was unclear.*

ADJUDICATION AND RESOLUTION

Alexander Lindsay was granted a pardon and an undisclosed amount in settle-
ment. His rights were restored but not the life he knew prior to the time of the
crime. There were other viable suspects to the assault on Pamela Lindsay. One
such suspect was described by the author Anne Rule in her book, *A Rage to Kill*.[6]

ADDITIONAL NOTES

After the final report of the second inquiry was published, slides were received of
transfer patterns on the bed sheet of the Lindsays' son. The imprint (Figure 6-17,
see page 76) confirms that the weapon was used horizontally rather than overhand.
Why it was laid on the child's bed, before moving to near the front door where it was
found, is unknown. Note the lack of blood on the tip of the weapon. The bloodstains
show the area on the weapon that contacted Pamela's face. The conclusion is that
the tip was caught by the wardrobe during the beginning of the attack, thus causing her attacker to shift to a horizontal position for further blows. This probably

---

[6]Rule, Anne. (1999). *A Rage to Kill*. Simon & Schuster Inc., New York, 187–238.

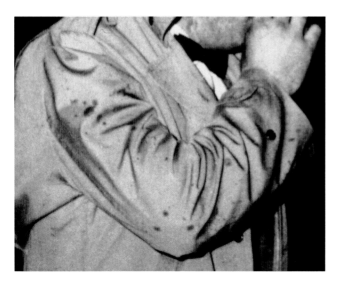

Figure 6-14

News file photo of Alexander wearing his jacket. The blockage area can be associated with his bending elbow holding the cigarette. Note color variations of stains on the jacket as well and lack of them on the upper cuff area.

saved her life, and Alexander's too (had he been convicted of murder in Australia instead of just assault).

WHAT CAN BE LEARNED FROM THIS CASE

At a conference held in South Australia, several police officers approached the author with statements that "Lindsay had blood on him so he must have done it." The answer for all was that although blood was found on the suspect no consideration was given at the time to "how he had acquired the stains." This is probably the most important lesson from any case that reverses itself later. A well-established criminalist and author stated, "the husband did it, the husband always does it." The error in this is the word "always." The attack on Pamela Lindsay appeared to illustrate a profound hatred of either her personally or of women in general. The lack of continued attack on the boy tends to support a specific hatred of women. If Alexander Lindsay did such a violent crime, something should have existed in their relationship to confirm such strong feelings. Nothing was ever shown of a motive or a predilection toward women of such strong emotion. In fact all indications from interviews of people who worked with and knew Alexander confirmed he had an old world attitude toward women, that they belonged on pedestals protected and cared for by men.

Another point that can be appreciated comes from the performance, or lack thereof, by this author. Initial contact with the defense barristers and Crown officers showed that she knew what she was doing. First appearance in court, however, provided spectators with a view of the worst performance from an expert imaginable. By the fifth day of the hearing, the expert's performance improved to court expectations, but not before the judge and barristers had cause to question their expert selection. On the fifth day the police Crown barrister asked of the witness if the stains identified "were medium velocity impact spatter."[7] Such terminology was not used previous in this inquiry. Before the witness could point out that medium velocity could apply to any of the patterns identified such as bludgeoning, impact, cast off, and arterial damage, the judge interceded and asked of the barrister if his query regarding medium velocity impact spatter were in reference to arterial damage, cast off, or impact

---

[7]Report of the Inquiry Held under Section 475 of Crimes Act, 1900 into the Conviction of Alexander McLeod-Lindsay, October 1969.

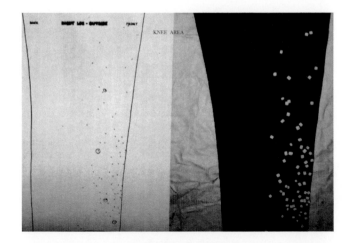

Figure 6-15

The pants photographs show where spots were found. A misconception was that the whole pant leg was included. The view is of just a little above the knee to the cuff.

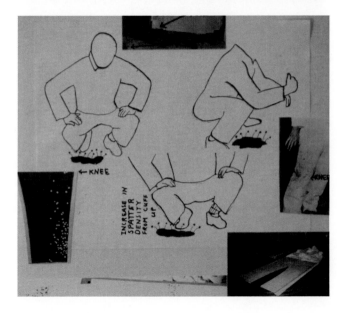

Figure 6-16

Suggestions for how Lindsay could acquire spatters on his pants were demonstrated. The court preferred the idea that he acquired "bouncing blood" from standing next to his wife at the hospital.

dynamics. The judge had learned the difference in dynamic acts even from a poor witness.

This represents a possible future in BPE testimonies. It was and may still be found that the discipline is applied as a comparison between curriculum vitae of the experts. One expert says the pattern is MVIS consistent with a beating and another expert says it is MVIS consistent with respiration (exhalation, expiration). The jury is left to choose between the qualifications of the experts rather than truly understanding the evidence itself. In the Lindsay second inquiry the judge was instructed on how to identify different dynamic patterns throughout eight days of testimony, so that the evidence itself was showcased and successful in assisting the trier of fact, even with a poorly performing expert.

Figure 6-17

Transfer patterns of jack bit on boy's bed sheet showing very little blood on the tip.

Ironically, bloodstain patterns often are compared with fingerprints, yet fingerprint testimony is presented in a stepwise, factual manner. The loops and whorls, number of ridges, breaks, and bifurcations are tallied to illustrate to the trier of fact how identity is concluded. This same approach is not applied to bloodstain patterns except in regard to size of individual stains, even though BPE often is lumped with fingerprint evidence. It has long since been determined that individual stain size depends more upon the degree of force within the dynamic acts than on the specific event type of act. The size of a stain alone cannot be used to identify a criminal act. There are too many variables involved.

The lamentable tragedy of Alexander Lindsay shows how much better it is to be sure of your suspect at the first trial. Too often overworked and underappreciated law enforcement officers finalize a case with limited evidences and move on with the belief that if the person is innocent some clever attorney will get them off. Unfortunately it isn't that simple. Once convicted or acquitted, the efforts to correct a person's life or justice becomes immeasurably harder, if not impossible. The use of experts based upon their resumes rather than the opinion of facts of any given case may tip the balance unfairly, and should be challenged by responsible scholars in law and forensic science.

Figure 7-1   View of bathroom after victim of GSW to the head was removed.

CHAPTER 7

WHO WAS THE SHOOTER?

Case 3

BACKGROUND

Three males (A, B, and C) and two females (1 and 2) were partying in a motel room. The females were girlfriends of males A and B. Male C demanded time with Female 1 and an argument ensued. The three men went into the bathroom, closed the door, and a gunshot was heard. Male B exited the bathroom, closing the door behind him, and immediately left with both of the females. Male A remained in the bathroom for a short period of time then exited the room, leaving his shoes by the bed, and the "please make up this room" sign on the unit door. Maids entered the room the next morning, stripping the beds before opening the bathroom door, where they discovered the body of male C. Law enforcement patrol officers responded first then called for detectives and a crime scene technician.

Fast food containers were found in the room, which led to a screen of the surveillance cameras for stores carrying the brands represented. The victim and one of the females were seen on one mini-mart tape. The female was located and encouraged to relate what took place at the motel. She stated that she heard the gunshot but never saw the gun nor did she know who possessed it. She identified which of the two male survivors came out of the bathroom first, and that this man related "A just shot C." The females and male B left the scene and did not see anything further but confirmed that male A must have stayed at the scene a short time after the gunshot.

BLOOD SOURCES

The major blood source was the single gunshot wound to the head (GSW/head). Head wounds tend to bleed profusely, providing blood for drip cast offs (LVIS, passive, gravitational). The injury was a through and through, exiting back of the right temple, but the exit lacked sufficient force to carry further. The projectile (bullet, missile) was found on the floor to the left (facing) the commode, or the victim's right as he faced the closed bathroom door. The bathroom was divided between a tiled portion, from tub past commode to counter edge, and a carpeted portion, from counter edge to the wall. The body was found with the head on top of carpeting with legs extended over the tiled portion. Blood had accumulated in a volume (pooled) stain extending underneath the carpet, around the head and upper body. The portion under the carpet remained wet for some time after the shooting.

BLOODSTAIN PATTERNS IDENTIFIED

A variety of patterns were found (Figure 7-1), although no clear pattern was seen to identify the location of the gunshot wound to the head. A suggestion for locating an **area of convergence** was present as a small (less than 3-inch diameter) scattered arrangement of medium-sized (visible stains, no microscope examination) spatters on the wall behind and above the commode tank (photograph unavailable). The bullet entered and exited oily, thick hair. It was possible that conditions were not favorable for entrance or exit wound, gunshot distributed **entrance wound spatter** (GDIS, HVIS, back spatter, blow back, forward) to be recorded.

On the floor was a large whole pattern of drip cast offs (passive stains, drip trail, gravitational stains) composed of several linearly arranged, round individual stains in a wide arc whole pattern. These correspond to blood dripping from tendrils of hair on the victim's head and follow a course from seated on the commode leaning to his right, rocking to the left, and to the final resting place on the floor.

The drip cast offs flowed so heavily that blood into blood resulted as the head was moved to the side. This is seen in the spatters projected from the floor up the base of the commode (Figure 7-2). More secondary spatter from blood into blood was seen on the back of the door facing the commode (see Figure 7-3). Two partial (area blocked part of the time not continuously) **blockage transfers** (voids) were seen on the right side (facing) of the door. There was also a partial blockage seen on the floor corresponding to the path of the door opening (Figure 7-4).

Figure 7-5 shows a suggested act of the shooter holding the victim while the second man exits the bathroom. Dripping from the head wound created satellite spatters, which outlined the events following the gunshot.

Hair swipe transfer (Swipe) is seen on the counter top in a position consistent with the drip cast off patterns on the floor (Figure 7-6).

Footwear impression transfers are seen in two places. One is at the edge of the volume (pool) on the carpeted floor, and the other is on the tiled floor over secondary spatters from the drip cast offs (Figures 7-7 and 7-8, see pages 83 and 84).

Area of convergence: An area determined from the directions of travel of several impact distributed spatters.

Entrance wound spatter: Blood drops distributed from an entrance wound.

Blockage transfer: A stencil effect where spatters surround an identifiable obstruction.

Figure 7-2

Rapid blood dripping, drip cast offs (LVIS, passive, drip trail) splashed upward onto the commode. Small-sized stains were more from rapid bleeding than from height of fall. Note size of each floor tile is 1 inch square. The size of each stain resulted from multiple drops recording as one bloodstain.

Figure 7-3
Shadow blockage transfer is seen on the back of the door. Someone blocked the door, then left while blood was still dripping into blood from the front of the commode, so that some stains were on the back of the door, but frequency was less in two places.

Figure 7-4
Satellite spatters from blood into blood cover the bathroom floor but include a shadow blockage where the door was opened while the victim was dripping. Someone had to hold the victim while someone else opened and closed the door.

It is noted that no smudging or distortion occurs to the spatters under the shoe impression, which is interpreted as the primary spatter stains being dry before the shoe print was added. This in turn suggests that the shoe print was a part of investigative transfer (**contamination**, secondary transfers) from a first response officer later.

Swipe transfer patterns were found in relation to the victim's hand with suggestion of clot initiation on the garment found there (**PABS/clot-I**). When clot forms on open weave fabric, the retraction process tends to draw the material up into wrinkles rather than pulling of cells away from the garment. This shows clotting occurred after blood pooled, in this case in the hand. The directions of the swipe (spreading and depositing blood along the movement) suggests the hand was under the body and acquired blood from the volume accumulation (pooling) before being pulled out and the garment added.

Contamination: Changes in evidence that results from actions of bystanders and rescue teams.

Swipe transfer pattern: A moving transfer pattern where blood is transferred to a non-bloodied recording surface.

PABS/clot-I: A bloodstain where stages of clot were reached before distribution.

Figure 7-5

Shows a suggested act by the shooter while the second man left the bathroom.

Figure 7-6

A swipe stain on the counter edge shows the hair was bloodied and that it brushed against the counter as the victim was allowed to fall to the side onto the carpeted area of the bathroom.

OBJECTIVE APPROACH TO INTERPRETATION

Partial blockage, shadow blockage transfer,[1] shows that the door was closed at the moment of the shot but opened and closed during rotation of the body (i.e., during secondary spattering (satellites) from the drip cast offs (parent drops) and pitching forward of the victim as he was moved to the floor. The person moving the victim to the floor was the shooter (male A), who left the bathroom after male B. The person partially blocking the door was standing in front of the victim, who was seated on the commode lid facing the door, when the shot was fired from the victim's left.

Since the bullet traversed the brain, thus making him incapable of voluntary acts, someone had to hold the victim to rotate the body. The location of the beginning of the drip cast off arc shows that the victim could not have held himself that far out of plumb over the floor by himself. This confirms that someone held the body as it was rocked to the left and allowed to collapse on the floor. No patterns suggest that the body pitched completely forward immediately following the shot. Someone caught him and guided his body to the floor. The person standing in front of the door exited, so was not the person moving the body to the floor. This indicates that the person holding the victim, while the second person exited, was the shooter. There would not

---

[1]Wonder, *Blood Dynamics*, 85.

have been time for the two men standing to change places before the victim pitched forward.

The shot was delivered right-handed. Although a left-handed person could deliver a shot with the right hand, the position would be uncomfortable and awkward to align entrance and exit as happened. The combination of entrance wound, head position, and location of the bullet after exit, with rotation of the body to the floor, shows the victim was facing the door at the moment the shot was fired into his left temple. This indicates the presence of a second person in the bathroom who was standing in front of the victim, i.e., the victim was facing male B when male A shot him. Male B was left-handed and male A was right-handed. Male A was the shooter.

Shoe prints were added after all the drip cast off and secondary spatters on the tile floor were dry. The blood source was clearly the volume under the carpet, which was still wet when first line patrol officers responded. The footwear impression evidence was classified as **investigative transfer** (contamination).

The interpretation of the patterns positions each of the participants in the bathroom during confrontation. The victim was seated on the commode, assailant A was to his left and assailant B was standing in front of him. The gunshot was fired, entering behind the left ear and exiting without force behind the right temple. The victim began to pitch forward and was caught by the shooter as the man standing in front of him left the bathroom. If the victim had completed his fall, his body would have blocked the exit from the bathroom. It is suggested that the shooter caught the body in order to prevent blocking his exit. Since the shooter was holding the victim the second man took the opportunity to exit first. This added corroboration to the female's story of who left the bathroom first, thus who did the shooting was concluded to be the second male exiting the bathroom.

Testimony was prepared to confirm the presence of three people in the bathroom. One left after the shot with the person remaining in the bathroom having to be the shooter. The shot was delivered right-handed. It was accepted and noted that right-handed people may shoot left-handed and *vice versa*, so the handedness was deemed of minor value at first.

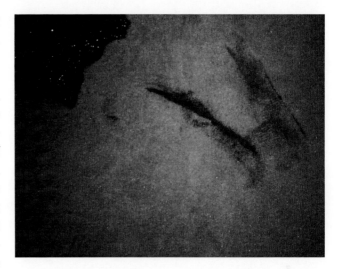

Figure 7-7

Investigative transfer, first line officer stepped on apparently dry carpet and forced blood underneath to surface. This became a blood source for a subsequent shoe print.

Investigative transfer:
When investigative officers cause changes to bloodstain patterns.

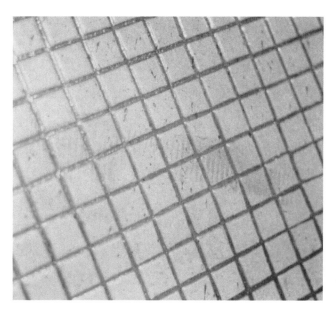

Figure 7-8

Note that spatters under the shoe print are not distorted. This supports the claim that the spatters were all dry when the print was made, i.e., not by the suspect wearing shoes.

ADJUDICATION AND RESOLUTION

The jury received the case a month after the bloodstain pattern expert testified yet credited the testimony with helping resolve the questions of guilt and degree of guilt. It was brought out during trial that one defendant was right-handed and one was left-handed. After this was brought out the right-handed defendant attempted to write with his left hand. It is unknown whether or not the jury saw this and/or was affected by it. It did suggest guilt to others who observed the act. The jury found the testimony regarding handedness more important than the original investigators credited it.

This illustrates the power of the evidence in resolving a charge, and also indicates influence with the trier of fact. The jury was impressed but the judge in this case was also influenced so that he forgot to fully instruct the jury. The choice was given to them that they could find the defendants guilty of murder in the first degree, second degree, and manslaughter. The option of finding the defendants "not guilty" was not given in court. Because of California's lengthy appellate process, by the time the case was sent back for retrial, suspect B, who was found guilty of manslaughter, was out of prison and stating that he'd claim to have fired the shot all by himself. The district attorney decided to not retry.

THINGS TO BE LEARNED FROM THIS CASE

Bloodstain pattern evidence can leave strong images with the jury that can help resolve a case long after the expert has left the stand. Right- and left-handedness can be important even when investigators feel that it is insignificant. Standard interview questions should include requests of all interviewees, witnesses as well as suspects, regarding whether they are right- or left-handed. Sometimes unsuspected witnesses interviewed become suspects later. Many times the issue of left-handed versus right-handed enters the case much later. It saves time to have that information available if and when the investigation focus shifts.

Whenever you have two defendants, be aware that retrial after appellate review may be unreliable. The best bet is to do it right the first time.

Figure 8-1   Two rounds identified fired from the off-duty officer's .45 into the car.

CHAPTER 8

TRAFFIC SALUTATIONS IN AMERICA

Case 4

BACKGROUND

A man was speeding along the fast lane of a freeway in a compact-sized vehicle in clear daylight. He encountered an unmarked full-sized van, also in the fast lane traveling at the posted speed limit. The man in the compact flashed the headlights, honked the horn, and then bumped the van's rear bumper. The van driver refused to relinquish the lane, continuing at the same speed. The compact driver pulled to the slow lane, passed the van, and saluted the van driver by waving a .38 revolver. The driver of the van waved back with a .45. The two men then continued at high speeds, weaving in and out of traffic, exchanging shots, estimated at two rounds each.

The van driver was an off-duty law enforcement officer in his personal vehicle. The impatient speeder was in a borrowed car. The officer realized that shooting was not in the public interest and stopped pursuit, but after arriving at his duty station alerted all agencies in the area that the occupant of the compact might be seeking medical attention. The car driver headed for an emergency room. He had sustained an injury to his arm. Although the injury was not life threatening, it was bleeding heavily and the limb probably went numb. Law enforcement officers found him soon after admission to a local hospital.

The car driver was charged with reckless driving and evading an officer, although he wasn't aware that he was having a shoot-out with a law enforcement officer, nor that had he stopped he would have been arrested on the spot. He in turn charged the officer with shooting him in the arm. The officer admitted discharging two rounds (Figure 8-1). The man denied shooting back, although two rounds were missing from his firearm. Considerable blood distribution occurred.

BLOOD SOURCES

In some of the cases listed in this section, specific blood sources are indicated after examining bloodstain patterns; at other times the blood source is verified before looking at the BPE. Medical reports are usually much more detailed than autopsy reports regarding injuries to specific blood vessels, which would

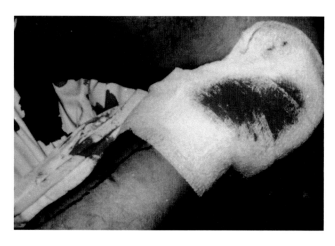

Figure 8-2

Injury to the deltoid muscle with arterial breach identified by ER staff.

provide sources of blood for distribution. This was important with the Alexander Lindsay case in Chapter 6, and also with this case. When a living victim is involved, it is essential to examine the medical records. Such is not always done because confidentiality laws surrounding medical records make obtaining them more difficult. Experience, however, has shown they are worth the effort.

The medical report listed a gunshot wound to the arm muscle, nicking the deltoid artery (Figure 8-2). Arterial damage occurred leading to a variety of pattern types after the initial injury: arterial gush (streaming), arterial spurt, blood into blood, and volume accumulation (pooling). Since these can be misidentified in the absence of knowledge regarding injury to arterial blood vessels, it is best to have the medical information before concluding the pattern identifications even though arterial blood distribution often is recognized before medical reports are available.

BLOODSTAIN PATTERNS IDENTIFIED

As noted earlier, a variety of patterns can result from a single arterial injury. Three main areas of bloodstains were found:

Figure 8-3

T-shirt worn by driver exchanging fire with off-duty law enforcement officer.

- T-shirt worn by car driver (too much contamination to be of use for BPE)
- Left side of the driver's seat (front and side), including the side of the head rest
- Driver's side door panel

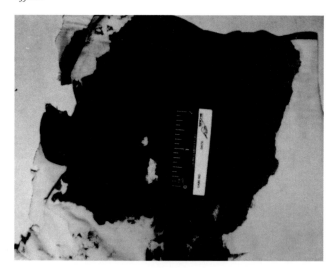

The T-shirt was not examined by this author. Photographs documented heavy, volume, blood staining around two holes of the shirt, which had been cut to remove from the patient (Figure 8-3). This was sent to a forensic lab for examination of residue, bullet wiping, and correlation with ammunition from each gun. Fortunately, the garment had been removed without cutting between or through the bullet holes.

The car was examined. A pattern was seen on the side of the driver's seat, which was a wide V shape based on alignment of outermost stains (Figure 8-4). The V shape contains medium-sized stains in a parallel arrangement.

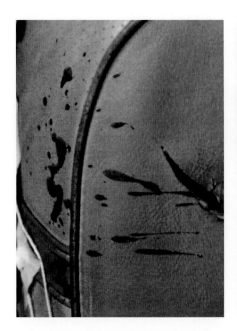

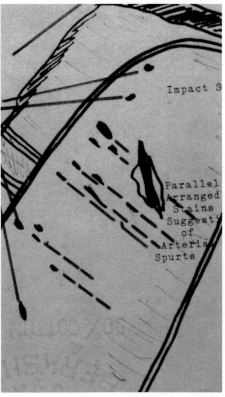

Figure 8-4

The left side of the driver's seat. V arrangement on edge of whole group with parallel, uniform-sized stains between are suggestive of GDIS and arterial spurting together.

Gunshot Distributed Impact Spatter (GDIS): A pattern which occured from firearm discharge into a blood source.

Figure 8-5

Multiple patterns overlapping: GDIS, arterial, volume, and blood into blood.

The size ranges were uniform with two different sizes noted. The whole grouping suggested overlapping dynamics involved in the blood distribution. The two events taken together that would explain the alignment were **gunshot distributed impact spatter (GDIS)** and arterial damage.

The door panel, shown in Figure 8-5, provides an excellent example of arterial damage composite bloodstain patterns in addition to an impact spatter arrangement suggestive of GDIS/entrance wound (HVIS). Blood drops distributed by gunshot are traveling at high velocity. This was noted by Dr. Paul Kirk as influencing the shape of the final stain to be more like exclamation points (i.e., more elongated). This can be seen in stains near the panel shelf under the driver's side door window. Drawing lines from two different groups of spatters, medium-sized stains versus the streaks, to their respective convergences locates two areas. One convergence is nearly a point and the other a much larger area. The point is horizontally located to the right of the larger convergence area (Figure 8-6). This suggests horizontal alignment of entrance and exit wounds from inside to outside the vehicle.

On the armrest, large ovals, noted earlier, in a parallel arrangement with directions of travel toward gravity are classified as arterial spurts, outlined in Figure 8-7. These lead to a volume stain (pool) on the armrest, which shows rapid bleeding forming a blood into blood (B/B) pattern. The projection of secondary drops from B/B was recorded on the door panel just above the armrest.

Figure 8-6

Strings locating areas of convergence.

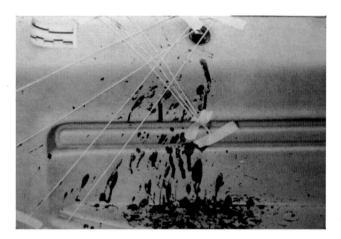

OBJECTIVE APPROACH TO INTERPRETATION

The greater uniformity and two size ranges identified this pattern as a probable combination of impact and arterial dynamics occurring simultaneously. This would be consistent with a gunshot to the deltoid artery with entrance wound aligned as inside to outside, or right to left with respect to horizontal position of the victim's arm. Round stains were seen on the front left side of the driver's seat. These did not suggest arterial damage because of the random sizes and distribution, round arrangement of spatters and roughly direct contact directions of travel (Figure 8-8). The classification suggested was as gunshot distributed impact spatter/exit wound (HVIS, forward spatter). Because this pattern was a direct (90 degree) contact from the drops distributed, only a suggestion can be made as to the event creating it.

It was concluded from the positions of the bloodstain patterns that the car driver shot himself while evading the van. The officer driving the van claimed to have discharged two bullets at the car. One bullet was found in the driver's side door and a second bullet entered the driver's open window, entered the head liner, furrowed along the top of the car and exited the passenger window, which was shattered as a consequence (Figure 8-9). The driver's side window was apparently down during the exchange but rolled up after the car driver was shot.

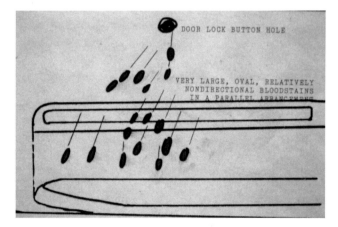

Figure 8-7

Arterial ovals outlined with clear plastic overlay.

The T-shirt evidence from two tests was presented, confirming that the bullet came from the car driver's weapon not the officer's. The examination by standard distance and amount of residue estimation provided results that the shot was fired from 3 feet away. Not only did the distance of the shot appear to contradict the conclusion that the victim shot himself, but caused some doubt regarding validity of the bloodstain pattern interpretation.

Two reasons were given for apparent discrepancies. The deltoid artery bled copiously, which would clean the wound and T-shirt of residue. It has been shown in research, however, that residue from firearms does not readily wash off.[1] A criminalist attempted to disprove the BPE by being driven along a freeway and dripping blood from a dropper into a tube. Butcher paper was used as a backdrop and later presented as proof that spatters on the seat could result from wind eddies. The results of the experiment (photograph unavailable), however, showed the dynamics to be flicking cast offs (cane-shaped whole pattern of large round dots) at the paper more than wind distributed patterns, which were then evaluated on the basis of spatter sizes alone. By contrast the case patterns were organized into separate dynamics overlapping with some forceful drop distributions identified.

Figure 8-8

Exit wound (forward spatter) GDIS on driver's seat back.

The second and most likely explanation for apparent discrepancies also provides information for future investigative leads. The road upon which the high-speed chase occurred was rough, and patched frequently. The car driver shot right-handed out of the driver's side window, crossing his right arm to his left side. To do this he steered with his left hand only. It is a reasonable suggestion that the gun was not steady in his right hand and that he attempted to rest it on his left arm, which was steering on a rough road. If the gun barrel jumped at the shot, the muzzle blast could encounter an area of the T-shirt offset from the bullet entrance enough to suggest a wider gap between gun and bullet hole when compared to stationary target tests. This explanation has two confirmations, experiments with the behavior of firearms[2] and experiments with a spinning firearms target (Figure 8-10), where the muzzle blast is separated from the bullet entrance.

Figure 8-9

One round accounted for from off-duty officer's firearm. Passenger window shattered.

The investigation application is that if two areas of convergence are found and associated with one entrance wound, the bullet entrance opens a blood source and distributes measurable spatters while the explosive muzzle blast that follows atomizes the blood and leaves mist and exclamation marks. This could provide a method of identifying rapid action at the moment of a shot.

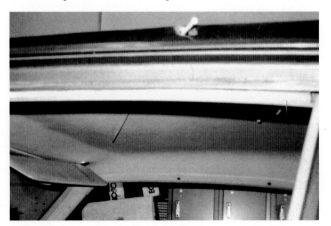

[1]Pizzola, Peter A., and De Forest, Peter R. (1997). A Recommended Protocol for Carrying Out Gunshot Discharge Residue on Garments. Oral presentation for the 49th Annual Meeting of the American Academy of Forensic Sciences in New York.
[2]DiMaio,Vincent. (1999). *Gunshot Wounds, 2e*. CRC, Boca Raton, FL, 49.

Spinning Target Counterclockwise

Figure 8-10

Spinning target exhibit (counterclockwise spin): The characteristics of a pattern often recognized as HVIS results from the explosive muzzle blast hitting blood exposed after a bullet entrance. No blood is available until after the bullet opens the wound. Note that the bullet forced the bloody sponge into the spinning target before entrance. Device is discussed in Section V, with research reproducibility.

ADJUDICATION AND DISCUSSION

Only traffic and weapon violations were brought and settled out of court. An attempt to sue the officer and department was unsuccessful since it could be proven that the car driver shot himself.

THINGS THAT CAN BE LEARNED FROM THIS CASE

This was an excellent example of composite bloodstain patterns where impact (HVIS), arterial damage, and blood into blood all overlapped. Instructors can use a clear plastic overlay on the figures to help students identify individual arrangements within the composite on the driver's side door.

A caution can be appreciated here also: gunshots are composites themselves[3] and they need to be considered within the context in which they occur. Stationary residue techniques must be interpreted with care when applied to very dynamic situations. If the victim was falling at the moment a weapon discharged, the residue analysis may indicate a greater distance than what actually occurred. The best way to approach this is to be aware that stains that are elliptical and measurable are more reliable for locating the origin of the bullet entrance. Bloodstains that are exclamation mark shaped or simply streaks may have been projected by the muzzle blast following the bullet, thus represent the location of the gun barrel and recording material a fraction of a second after the gunshot. Knowledge of the two would provide information of the true dynamics of the shot.

Reconstruction experimental design must take into account much more than reproduction of spatters of the same size as crime scene evidence. Bloodstain patterns at crime scenes frequently overlap, as well as spatter size ranges from considerably different dynamic acts. Therefore, the acts themselves cannot be identified from the size of stains alone. How the stains are grouped, the spatters within the group, directions of travel, their size range distribution, and their positions and density within the whole pattern must be examined for comparison of three-dimensional events, in crimes and reconstruction of crimes.

---

[3]Wonder, *Blood Dynamics*, 130.

Figure 9-1   A body was found on a closed vacation resort cabin porch.

CHAPTER 9

THE BODY ON THE PORCH

Case 5

BACKGROUND

During a late afternoon in early spring, a regular mountain bar crowd was drinking with the usual arguments. The locals included some ex-convicts as well as other rough survivors. An argument started between two patrons. The problem revolved around one man throwing his arms around another man, who had been raped in prison, and attempting to kiss him on the lips. The ex-con reacted, announcing to all present that "if you ever try that again, I'll stick you."

A week later, the same crowd was gathered and the same two men began the same argument. This time they left the bar to continue their discussion outside. A witness, who was aware of the previous confrontation, followed them. He stood across the yard drinking (position of glass in Figure 9-2), listening to the conversation between the two men. It was late enough in the afternoon that long shadows covered the ground but sound carried well. The witness claimed to not actually see everything that transpired but heard all the verbal exchange. The affectionate victim fell to the porch behind a short wall, while continuing to speak to the ex-con. The exchange was something like "why'd you do that for? I just want to be friends. Come on now let's just be friends." The ex-con then returned to the bar and left soon thereafter. No one checked on the victim. The witness assumed he'd fallen down drunk and would get up and go home soon.

Figure 9-2
Position of the glass locates where the witness stood and listened.

The next morning the victim's body was found apparently where he fell the previous late afternoon (Figure 9-3). The ex-con claimed the victim must have been attacked by someone else after he left because the victim was alive at that time, as verified by the witness. The bar patrons also confirmed that the ex-con had not returned to the cabin area.

Figure 9-3

The victim's body, as found, as he fell the evening before.

Attorneys often want the answers to specific questions, rather than full reports. In this case the questions were relative to an upcoming trial:

1. Do the bloodstains corroborate what the eyewitness claimed to hear?
2. Are the bloodstains consistent with the victim dying as found?
3. Is there any information that would tend to exonerate (confirm guilt of) the client (suspect)?

BLOOD SOURCES IDENTIFICATION

This incident took place in a rural community with the autopsy performed by a clinical pathologist. In this case both the preliminary investigation and the autopsy were conducted in a manner that suited the application of bloodstain pattern analysis. The victim had three wounds that would be expected to contribute to blood sources for patterns:

1. A cut was found on his right forearm which was not life threatening.
2. A stab wound was identified on his left side, which followed a tract that missed all vital organs.
3. A stab wound was located in his stomach with tracking through the descending aorta.

The last injury resulted in sufficient internal bleeding to be listed as the cause of death from blood loss. Because the descending aorta is deep internally, we would not expect to find arterial damage patterns externally, yet death from this one injury can occur rapidly.

The victim was photographed thoroughly as found, then disrobed at the scene and more photographs taken. This is excellent for identifying flows and patterns associated with the body. Too often, the body itself is ignored and the only photographs available are from the autopsy after the body has been either contaminated by transport in a body bag or after they have been undressed and cleaned up.

BLOODSTAIN PATTERNS IDENTIFIED

There are several individual patterns, predominantly of two classifications:

- Swing and drip cast offs (drip trails, passive stains)
- Volume and flows (pooling, flows)

One series of swing cast off patterns is seen on the wall of the cabin (Figure 9-4). Another series is seen as the cane-shaped whole patterns, flicked drip cast offs, on the porch flooring (Figure 9-5).

A flow from the stab to the side ran down the victim's side, then abruptly changed direction to flow at a right angle across the victim's stomach and down his back (Figure 9-6). The ends of some of the flows were at volume (pools) stains on the porch floor. No smudges or moving transfers were noted away from the volume stains. One smooth moving transfer, later classified as a skin swipe, was seen on the stucco of the cabin exterior wall (Figure 9-6).

Figure 9-4
Swing cast offs on the cabin wall.

OBJECTIVE APPROACH TO INTERPRETATION

Combining the various cast off patterns suggests a **struggle composite**,[1] and would be expected from an injury capable of moving and flicking blood drops. This is consistent with the arm cut, but not with stabs to the side or the abdomen. The smooth skin transfer on the wall, again associated with the arm, also confirms that this cut occurred while the victim was upright and mobile (Figure 9-7).

The cast offs on the cabin wall show that the arm was swinging in an arc suggestive of defensive or fighting motions. This could only have happened after the cut to the arm, opening of the blood source, and before the victim fell to the floor. The flicks on the floor could have occurred immediately after the victim fell. Swinging the arm in an arc suggested by the wall cast offs could occur during attempts to hug the assailant and also momentarily expose the victim's side to being stabbed.

The flow down the side shows that the victim was standing upright when the side stab began to bleed, but soon afterward he fell to the porch floor. However, the victim was found as he fell the next morning, seen with the volumes (pools) on

Struggle composite:
A group of patterns which illustrate the dynamics during an altercation.

Figure 9-5

Drip cast offs suggestive of flicking arm after injury.

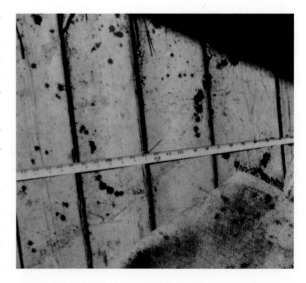

---

[1]Wonder, *Blood Dynamics*, 137.

Figure 9-6

Flows from the victim's side corresponded with pools on porch floor. Flow was from the side wound down the side and abruptly changed direction.

Figure 9-7

Skin wipe on the building. Note smooth texture and difficulty in determining contact and lift off. Skin smooths and spreads as it deposits (swipe) as well as when it wipes. This is further confirmation that injury to the arm occurred first and before falling to the floor.

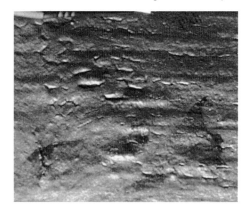

the porch floor under the flow from the side wound. Neither the stab to the side nor the arm would necessarily bring the victim down to the porch floor. A blow to the abdomen, tripping, violent shove, etc. might do the job. It is reasonable to suggest then that the abdomen stab that hit the descending aorta was delivered as the victim was standing and with sufficient force to shove him backward. This would mean that he was stabbed before he fell and that the wound which caused death was delivered just before he fell.

We can use the bloodstain patterns to sequence the three injuries: cut to the arm (struggle and swinging arms in an attempted hug or in self defense), stab to the side (flow starts downward), forceful stab to the abdomen, fall to the floor (with reversal of side wound flow), and death after the time period determined by the pathologist doing the autopsy.

The three questions can be answered by the bloodstain pattern analysis:

1. The bloodstain patterns corroborate what the eyewitness heard.
2. The bloodstains are consistent with the victim dying as found.
3. There is no information tending to exonerate the client (suspect).

ADJUDICATION AND DISCUSSION

The accused had adamantly claimed to have left the scene and someone else had found the victim and stabbed him later. When confronted with the bloodstain pattern analysis he accepted a negotiated plea.

WHAT WE CAN LEARN FROM THIS CASE

This case shows that bloodstain pattern evidence can assist in sequencing the delivery of injuries even if the fatal wound is not specifically identified in a bloodstain. Too often the focus is on spatters only. When they aren't found, indicated, or recognized the rest of the BPE is ignored. An objective approach to bloodstain patterns can provide investigative leads information regardless of the presence or absence of impact spatters.

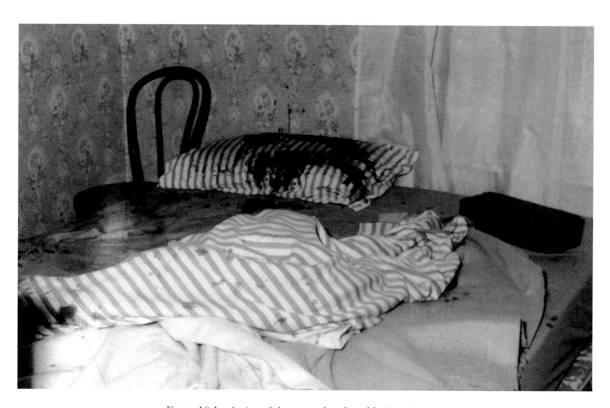

Figure 10-1   A view of the room when found by investigators.

CHAPTER 10

LIL' OL' GUY WHO WOKE UP DEAD

Case 6

BACKGROUND

An elderly man, with a history of alcoholism, was found in his home, dead from blood loss (exsanguination). It was hot summer, and he had no air conditioning, therefore windows were left open of an evening for circulation. The victim's body was found between the hall and the bathroom yet bloodstains were noted leading from the bedroom (Figure 10-1), along a hallway, into the bathroom. A suspect was arrested. Both the defense and the prosecution retained consultants in bloodstain pattern evidence.

The suspect admitted that he had entered the residence to burglarize. He claimed he heard the old man wake up and then quickly left the house. When informed that the victim was dead, the suspect claimed first to have left him alive, but had soon afterward told others in a bar about the old man's vulnerability. The implication was that one of the bar's patrons later had gone to the house and assaulted the old guy again. At some point the suspect changed his story and admitted to the assault but claimed it was in self-defense after the man awoke during the attempted burglary. An important factor was whether it was one assault, or two or more assaults with a lapse of time between, i.e., providing time for the burglar to escape if he was in fear of his life.

BLOOD SOURCES IDENTIFICATION

The victim had been beaten around the face and head. No mention of arterial damage was made in the autopsy report by a clinical pathologist, but the cause of death was listed as exsanguination blood loss. Two significant injuries were noted in the autopsy photographs, one to the mouth and jaw and one to the middle finger tip of the right hand. Either or both of these could feasibly involve small arteries. When asked later, the pathologist provided the location of the jaw as the source of arterial damage.

The victim's hair was bloodied but no injury was mentioned nor suggested in the photographs. This was classified as a prior stain, i.e., the hair became bloodied and provided an intermediary blood source for anything coming into

Inline beading: An in-tandem arrangement of spatters along the same line of travel.

Figure 10-2

Wet blood on autopsy tray.

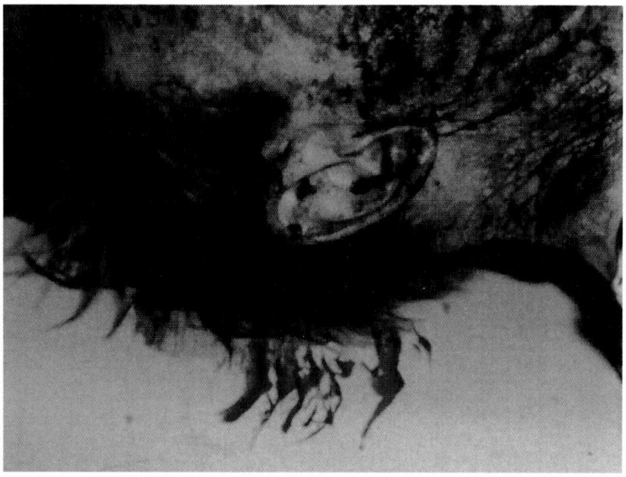

Figure 10-3

Three respiratory distributed patterns: over the bed to the right of the chair (above the pillow), to the left near the chair, and to the left of the chair next to the sheers.

contact with the hair before the blood dried. Blood was still wet on the hair at the autopsy, as noted from a moving transfer (swipe) on the autopsy tray (Figure 10-2).

Injury occurred to the middle finger of the right hand, resulting in the finger being mashed at the tip. Although not a commonly recognized source of arterial damage, the tips of the fingers do involve arteriole (very small arterial vessels) that lead into the capillary beds and on to the venous system. There is some pressure involved. This can account for weak arterial distribution from injury to the fingers, and arterial-like projection of drops distributed from squeezing punctures to the finger tips. Since the cause of death was blood loss, any possible source of arterial damage should be considered and checked with the autopsy physician. In this case, the clinical pathologist claimed that arterial damage to the finger would be minimal compared to that from the jaw.

BLOODSTAIN PATTERNS IDENTIFIED

Initial review of the scene focused on three patterns on the wall near the victim's bed (Figure 10-3). One pattern was a clearly defined oval whole pattern shape, with randomly distributed spatters of variable sizes, with directions of travel almost perpendicular but showing some orientation toward gravity (directly down). The second small pattern was to the left of the bed near a chair, and a third largest pattern next to sheer curtains (Figure 10-4). A few of the stains had inclusions suggestive of saliva, which would fit the classification of exhalation (expiration, respiratory) distribution.

On the window sheer curtain to the left was an irregular shaped fragment with residual fluid identified as PABS/clot. From the appearance of the fragment with aspirated residual fluid, the direction the clot traveled is suggested as from right to left, relative to the viewer (Figure 10-5).

Between the side wall sneeze pattern and the clot fragment on the sheer, on the bed was a large urine stain (Figure 10-6). The defense council associated this with the moment of death release of the bladder. Anyone dealing with the ailing elderly is aware that bladder release can follow nonfatal events such as sneezing, coughing, or incontinent behavior.

Laying on the right side of the bed, as viewed in Figure 10-1, was a large block of wood, shown in Figure 10-7. When the block was removed, no bloodstains were seen on the sheet beneath it, although heavy staining was seen on two sides with light staining on the other two sides. This suggests **PABS/dried**, i.e., the blood was dried, or at least absorbed thoroughly, on the block before it was laid on the bed. Efforts to locate another position where the block could have been left before being moved to the bed were not successful. This would have been a good use for enhancement techniques.

In the photograph of the block an indentation was seen with a wispy white/silver hair embedded under the blood (Figure 10-8). The block had been sent to a major crime lab, examined, and returned without the hair dislodging from the block. Arrangements were made to view the block at the department evidence storage later.

On the bed seeping into the mattress under the sheet in front of the pattern to the left of the bed (as viewed) was the urine stain. Roughly at the position of the urine stain, on the floor and leading to the door to the bedroom, were drip cast offs (LVIS, passive, drip trail) (photograph unavailable). On the bedroom door was a bloody hair transfer (compression, contact stain) (Figure 10-9). The hair acquired blood in a beaded arrangement, which transfers to a surface in a pattern described as **inline beading**.[1] The oil on the hair causes blood to bead up. When the hair comes into contact with a surface the beads along the shaft are recorded within the locks of hair.

The door frame showed several smudges near the light switch (Figure 10-10, see page 106). Next to this door frame, outside the bedroom, was a closet. On the front of the closet was a cascade of spatters that did not demonstrate an area of convergence. Instead, alignment was subtly shifting downward as if affected by gravity.

Figure 10-4

Third sneeze with the largest pattern next to the sheer curtains.

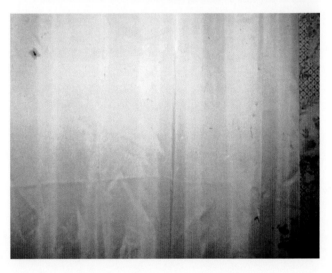

Figure 10-5

Clot fragment on drape sheers shows respiratory act was strong.

---

[1]Wonder, *Blood Dynamics*, 96.

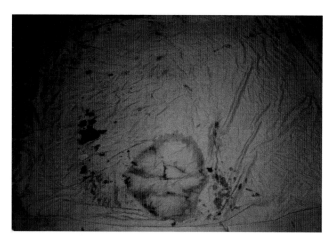

Figure 10-6

Urine stain on mattress corresponds to position of sneeze toward wall.

PABS/dry: A bloodstain where some degree of drying occurred prior to distribution (see previous page).

Figure 10-7

Close-up of block murder weapon.

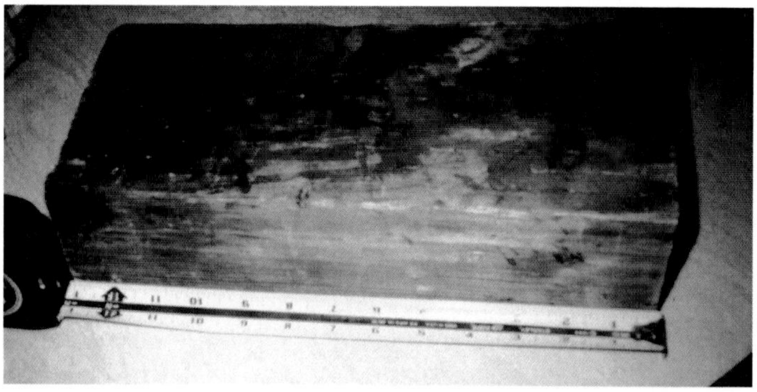

This was characteristic of cessation cast offs, i.e., a weapon with blood adhering was swung right to left and stopped abruptly at something near the bedroom door frame. The blood drops then separated from the carrier and continued travel right to left but pulled downward from gravity (Figure 10-11). The swing preceding cessation would have been from right to left, hitting the left side of the victim's head/face but terminating at the jaw with cessation cast offs continuing on to the closet door.

Direct transfers (compression, contact stains) were seen on the door frame between the bedroom and the hall. The pattern resembled four finger marks in blood. The four fingers of the right hand (determined from the arc of the finger joints and position of the little finger) were seen as normal-sized exemplars. On the wall leading from the bedroom to the bathroom were a continuing series of finger-like mark contacts, showing one tip larger and apparently mashed and corresponding to drip cast offs with directions of travel toward the bathroom (no photographs available). This suggests that the middle finger of the right hand was damaged between the bedroom doorway and the first series of prints on the hall wall (i.e., at or near the location of the closet door). The victim was alive and walking to the bathroom through the hall when assaulted a second time, the first time being where blood began to flow in the bedroom. The injuries to the hands were also suggestive of defense wounds, not aggression, for the self-defense claim.

In the bathroom were numerous arrangements all identified as arterial damage patterns. Large oval-shaped individual stains, lacking directions of travel, and arranged parallel with each other, were seen flowing by force of gravity alone. These were in the sink (Figure 10-12), on the tub (Figure 10-13, see page 107), and on the bathroom door next to the body (Figure 10-14). In all three areas blood into blood was seen on several horizontal surfaces, showing rapid blood loss occurred in the bathroom.

Later at the police department the wood block was reexamined. The wispy hair was noted within the heavy layer of blood at an indentation to the block. The hair was embedded in the block and also held there by dried blood. The original examining criminalist claimed to ignore the hair because both the hair and block belonged to the victim. The position of the hair within a marked indentation, and under a heavy bloodstain impact pattern, provided a context that was part of the crime. Bloodstains on the sides of the block showed convergence of an impact spatter (MVIS) pattern confirmed that the hair was part of the evidence of the crime.

OBJECTIVE APPROACH TO INTERPRETATION

The patterns seen on the walls above the head and to the right of the bed as viewed were identified as respiratory/sneeze (exhalation, expiration) patterns. The fragment on the drapes would require more energy of projection than a cough or wheeze would

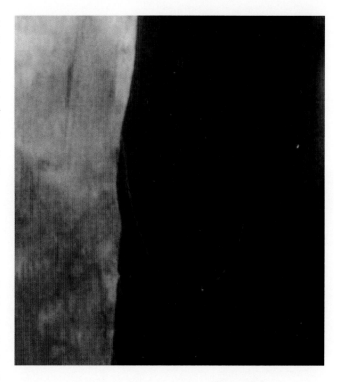

Figure 10-8

Close-up of wispy hair attached to block at impact area.

provide, but would be consistent with a sneeze. The voiding of urine could accompany a strong sneeze with the elderly. The victim was allegedly an alcoholic, sleeping off a drunk when attacked. The suggestion that the pattern on the wall was an impact spatter (medium velocity impact) pattern was contradicted by the shape (oval arrangement of the group of spatters) and location of the whole arrangement of spots (on the wall above the pillow). Injury to the victim's face to distribute the array recorded on the wall would require the blow to occur between the victim's head and the wall. If such a blow were delivered, it would also block recording of spatters. Pantomiming the act of hitting the victim with a large block shows the improbability of the patterns being impact in origin. The accused finally admitted to hitting the old man while he was still in bed. It was the subsequent blows that he tried to deny.

Figure 10-9

Simple direct hair transfer on bedroom door surface.

The bloodstain pattern evidence shows that there were at least two assaults, while the victim was on the bed and when he reached the bedroom door. It is possible he was attacked a third time at or in the bathroom. If so, the arterial damage patterns overlapped any indication of it.

A confirmation that the man was bleeding around the face while laying on his back was seen in the western style shirt in which he was found. Spatters are absent from the pocket flap (Figure 10-15, see page 108). Spatters are seen on the lowest edge of the pocket, but not under nor on top of the flap. If the victim were upright, the flap would

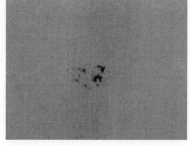

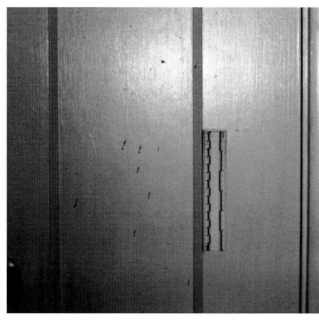

Figure 10-10

Fingerprints in blood at light switch.

Figure 10-11

Cessation cast offs on closet outside bedroom (close-up views not available).

Figure 10-12

Arterial damage patterns seen in bathroom sink.

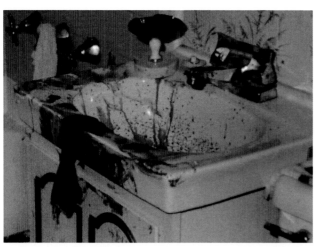

have been down and only the top side exposed. If the flap were up, the underside would receive blood. The flap was sticking up perpendicular to the shirt, neither laying flat normally nor reversed. The victim was laying flat exhaling blood that settled back on the pocket but only the tip of the flap.

The clot fragment indicates a delay in time between an assault to the victim and his attempt to rise. Blood in the nose or mouth could cause the need to sneeze, with sneezing responsible for urinating. The need to urinate could have induced the awakening man to head to the bathroom. Sequencing events provides the first assault to open a blood source at the bed, then a period of unconscious breathing, sneezing, sitting on the side of bed, sneezing a second time, then efforts to walk to the bathroom, reassaulted at the doorway causing the jaw artery to be breached and an injury to the middle finger of the right hand, stumbling back against the door with transfer of hair blood, then continued down the hall to the bathroom, leaning over the sink, falling to the tub, crawling to door, collapsing, and dying.

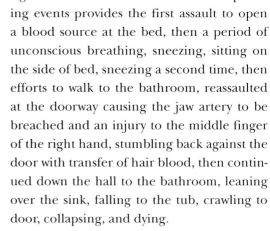

At least two assaults occurred, with the victim being unconscious between the two. This

Figure 10-13

Arterial damage bloodstain patterns seen on tub.

Figure 10-14

Final resting place of victim by bathroom door.

nullifies a self-defense claim. The force of the attack and confirmation of a second attack was corroborated with the imbedded hair in the block.

Most of the patterns involved were not offensive if presented as isolated photographs. The block itself seems a bit unwieldy if handled to avoid touching the bloodstains. A reasonable facsimile was constructed by photocopying the block. During photocopying the silver hair continued to adhere tightly to the area of heavy blood staining. The photocopies were attached to a clean block of the same size and shape as the actual weapon (Figure 10-16). This was available to allow the jury to examine the block, hold and handle it, and most important to use in different manners and angles without concern for blood staining. This would permit them to determine to their own satisfaction how it had been used in the attacks. The prosecuting attorney did not request use of the demonstration block in this case.

ADJUDICATION AND RESOLUTION

Information was provided to the author that the case pled out and no trial would occur. No other details were available.

WHAT WE CAN LEARN FROM THIS CASE

The main point was the importance of arterial injury. Two injuries suggested that arterial patterns could be involved, the jaw and the finger tip. It has been a

Figure 10-15

Pocket flap showed no blood spatters on top or under flap only above and below.

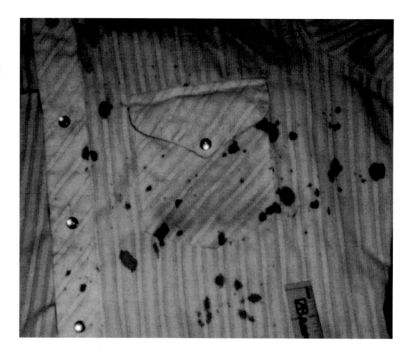

Figure 10-16

Mock-up of murder weapon for jury examination. Not used at trial.

policy of pathologists in the past that arterial damage is not necessarily mentioned if it is not involved in the cause or manner of death. With the amount of blood seen in this case and the cause listed as blood loss, arterial damage should always be mentioned in the autopsy. In the future it is hoped that listing of damage to arterial blood vessels will become a standard policy with all death investigations.

Another point learned from this case is that trace evidence is still alive and beneficial to the resolution of casework. The blood on the identified weapon was typed. No pattern interpretation was applied nor was the wispy silver hair noted or described. Today overworked and underappreciated criminalists may do DNA and ignore other evidence. In this case DNA would have added nothing of investigative importance to the case. The BPE clearly indicated two attacks, and the imbedded hair connected the block with directed violence rather than self defense. Whether the hair was of significance or not was up to the prosecuting attorney. The decision to use or not use evidence, however, could not be made if the information was not available. Presence of the hair at least should have been listed for the attorney's information.

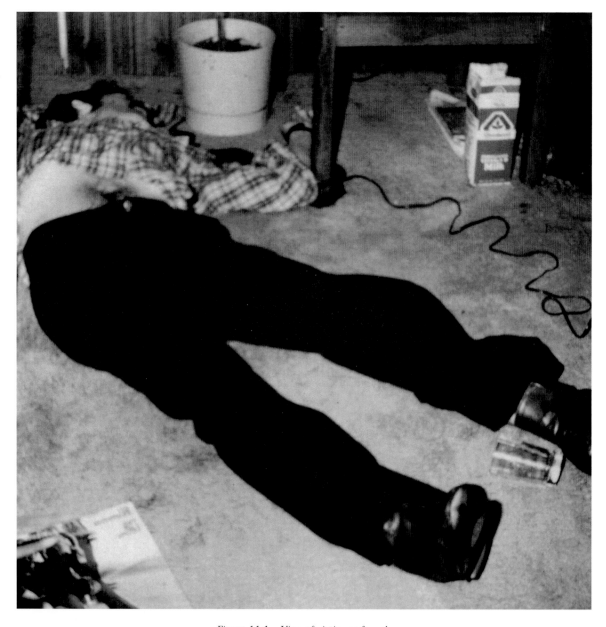

Figure 11-1   View of victim as found.

CHAPTER 11

SELF-DEFENSE STAGING HOMICIDE BY GUNSHOT

Case 7

BACKGROUND

Two men were rooming together in a small apartment in a city. One man (later the victim) was a known alcoholic with a reputation for being an abusive drunk. The other man (later the accused) was a known brawler with a long rap sheet of petty offenses, but no previous homicide accusations. The alcoholic was found dead from a single gunshot wound. The roommate, suspect, first claimed to be away at the time, however eyewitnesses placed him home when the shot was heard. The statement was then changed to self-defense, using the fact that the victim was known as an abusive drunk.

BLOOD SOURCE IDENTIFICATION

The injury was apparent from photographs, later confirmed from the topsy report, as a gunshot wound to the left side of the face below the left nostril. It was stated that the wound was from a 0.38 caliber bullet that entered below the nose and exited the skull, but not the scalp. The autopsy claimed the star-shaped wound to be close, near contact, range. The only source of gunshot distributed impact spatter would be entrance wound impact spatters (GDIS, blow back, back spatter, HVIS). No exit wound impact spatters (forward spatters) would be associated with the injuries as described in the autopsy.

The autopsy also noted that the left lung had over 100 cc of blood accumulation. This could provide blood for possible respiratory (exhalation, expiration) patterns. A volume stain (pool) was located above the victim's left shoulder. This could be a source for transfers, cast offs, **splash,** etc. Prior wet stains on hands might have been involved in arrangements of spatters in different shades of pink and red seen in the kitchen sink. These latter are discussed later.

Splash: An impact to a volume of blood.

BLOODSTAIN PATTERNS IDENTIFIED

The first review of the case was with photographs sent without autopsy or interview information. Although impressions from the photographs were later confirmed

Figure 11-2

An end table with various patterns overlapping was the initial focus.

with two visits to the crime scene, the photos provided sufficient information for the majority of the identifications and interpretation listed later. The crime scene showed the victim was on the floor in an apartment with his left arm along the edge of an end table (Figure 11-1). The primary concern of the prosecuting attorney was the array of spatters on an end table to the left of the victim's body (Figure 11-2). The specific request was for an expert who would say that the spatters were either HVIS (GDIS) or inspiration [sic] (expiration, exhalation, respiratory). Because the size of the spatters was well above the range taught in a traditional program to label a pattern as HVIS, no confirmation could be given. There were, however, many other patterns of interest, including a fragment of clot within the GDIS (gunshot distributed impact spatter) on the table.

The most important observation with the photographs was a flow and blockage transfer (void) outlining the victim's left eye. This was best demonstrated on the autopsy photo, before the body was cleaned (Figure 11-3), but also visible in the original crime scene view (Figure 11-4). The flow appeared to run from the stellate entrance wound beside the left nostril to, but not within, the left orbital space. On the right eyelid, however, was a single small spatter, also seen in a close-up of the scene photo. The stains were thus verified as on the body at the scene, and not defects from photography or contamination from transport to the morgue.

Two pairs of glasses were still present when the scene was visited weeks later (Figure 11-5). One pair was for reading and the other was nonprescription sunglasses. Each pair was folded in a different way, suggesting they were folded by different people.

Figure 11-3

Victim before cleaning for autopsy. Note outline of eyeglasses.

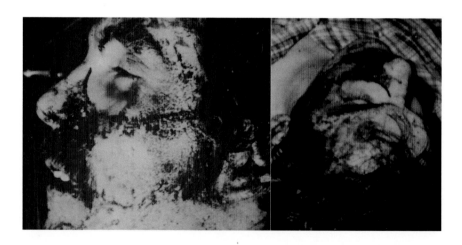

On the end table front left leg (or right side as faced, i.e., farthest from the victim's left side) were overlapping spatter patterns, requested to be the main focus of the BPA. One reconstruction with these was made by drawing lines on the photograph to show a *point* of origin in front of the victim as his body was found. From that it was assumed that he was shot while seated on the floor, propped up on his elbow listening to records, and immediately fell backward at the shot. Much of that scenario was contradicted with other evidence although apparently retained through trial.

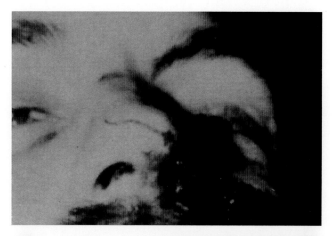

Figure 11-4

Verification that glasses stain was present on victim at crime scene.

On the opposite front end table leg (left side as viewed next to the victim) were a row of large oval stains, lacking clear directions of travel and overflowing straight down toward the floor. These were identified as swing cast offs from an event that occurred roughly in front of the table leg and involved heavy blood flow (Figure 11-6).

Under the table were medium-sized spatters with area of convergence consistent with (would draw to a small area of convergence) the spatters on the left table leg. The apparent area of convergence was toward the position of the victim's waist as he was found. The presence of a convergence rather than sizes of the spatters confirmed their classification as impact distributed. Near this area, next to the victim's waist as found, was also located a red cross arrangement seen in the photographs and later verified at the scene (Figure 11-7). This corresponded to other convergence areas and suggested the actual location of the gunshot to the victim's face. If this was true, the body had been moved after the shot.

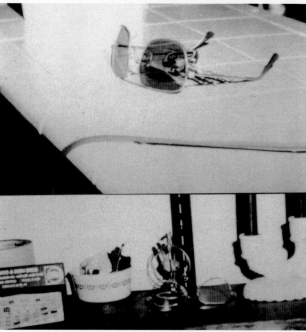

Figure 11-5

Two pairs of glasses were still at the scene when visited weeks later.

A drip cast off with secondary (satellite) spatters was seen on the palm surface of the left hand, yet the left hand is found under the end table. Drip cast off (LVIS, gravitational, passive stain) could not occur as the left hand was found (Figure 11-8). This also suggests that the victim was moved after the shot and the size of drips is consistent with the cast offs seen on the left front table leg.

The volume stain (pool) located above the victim's left shoulder was positioned in front of a scattered arrangement of medium and large-sized stains.

Splash: An impact to a
volume of blood.

This suggested **splash** (splatter) on the carpet surface. Placing an outline of the accused-size shoe sole over the stain showed stomping in the volume (pool) was a possibility (Figure 11-9).

A plant container was positioned against the wall and in alignment with the splash. The front of it had received spatters arranged in a parabolic grouping suggestive of spatters hitting the front and continuing to travel according to the path of least resistance[1] slightly down along the sides. The curve of the stains appears to converge at a point about 5½ inches from the floor. The victim's head measures 5½ inches from the back to the area of the gunshot. The resemblance to an impact spatter pattern (Figure 11-10) and height measurement led to its identification as impact spatters in this case. Experience with crime scenes, experimentation, and other photographs suggest that the pattern could also have been a mixture of spatters from splash and impact.

Blood transfers are seen between the fingers of the right hand. These appear to have transferred between fingers above the web spaces. Folding the hand to see if such an action was possible shows that joint transfers would be impossible but transfer from adjacent fingers would be a logical explanation (Figure 11-1, see page 117); i.e., the fingers were apart when exposed to a spattering event. This would be more consistent with a defense gesture rather than an aggression.

Figure 11-6

Spatters seen on the left front leg of the end table show predominant cast offs but a few impact spatters also (identified from their area of convergence, not size).

OBJECTIVE APPROACH TO INTERPRETATION

The victim's right hand was open with the back side toward his face at the moment of the shot, exposing surfaces to blood spatters. The hand closed at or soon after the shot transferring blood to adjacent fingers. Therefore, the hand was held up with fingers splayed, a defensive gesture, rather than as the fist of an aggressor.

---

[1]Wonder, *Blood Dynamics*, 86.

The autopsy report showed 100 cubic centimeters of blood in the left lung. The pathologist, however, claimed that the victim was incapable of breathing after the gunshot; therefore something opened a blood source prior to the gunshot. The presence of blood in the lung was explained as the victim took a dying breath. If the victim was incapable of breathing, a dying breath would also be impossible. There is a further problem with that assumption. Victims do not inhale at death, they exhale. This is the sigh often noted at death beds. In any case, 100 cc of cohesive clotting (fragment seen on table leg) blood would not be inhaled with a single breath.

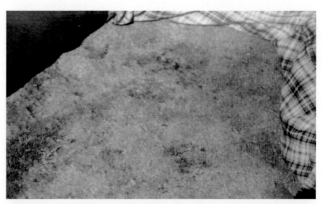

Figure 11-7

Red X on the carpet near the victim's waist.

Many years later a detective pointed out to me the importance of the cleaned autopsy photograph (Figure 11-12). The bruised lips and injury between the eyebrows confirms a blow to the face prior to death. The fact that a fight occurred before the gunshot was confirmed with the BPE outline around the eye. Detectives interviewed residents below the apartment of the victim and accused. The breaking up of ice in the kitchen could have served for a mixed drink (glass seen on floor by victim) or been used for a nose bleed from an earlier blow to the face (autopsy bruising between the eyes and on the lips).

Although not a bloodstain pattern, the twist in the victim's pant cuffs provides another indication that the body had been repositioned after the shot. One pant leg is twisted and the other is bunched up as if the legs were moved by someone other than the victim, and separately positioned.

While blood flowed, the victim was wearing a pair of eye glasses positioned on the face slightly askew. An experiment (Figures 11-13 and 11-14) showed that the time period necessary would equal or exceed 45 minutes to not flow into the orbital space.

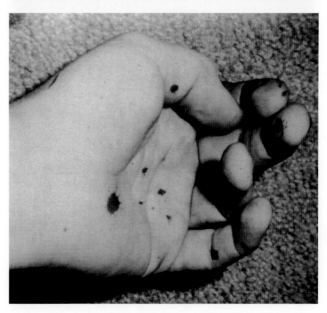

Figure 11-8

Drip cast off in the hand, which could not have acquired the drip as found.

This would be consistent with the victim laying on the floor unconscious with blood from a bloody nose flowing into his lung and up around his glasses. That the glasses had to be partially askew confirmed that a fight occurred before the shot and that the nose bleed could have resulted from the fight.

Another experiment was conducted with regard to the blood outlining a pair of glasses. The orbital space had been protected from blood distributed by the gunshot, yet a single spot was deposited on the eyelid. If the spot was from splash after the gunshot, a bigger drop and more of them would be expected. To test what happens when a gunshot causes the head to pivot, a device with a pin release was constructed (Figure 11-15). As a gunshot was delivered to a styrofoam head held in the device, it would pivot to a position duplicated in the photos of the victim (Figure 11-16, see page 119).

Figure 11-9

Shoe outline in pool. Spatters seen above victim's head probably include splash.

Figure 11-10

Blood spattered planter was a focal point to an inexperienced investigator.

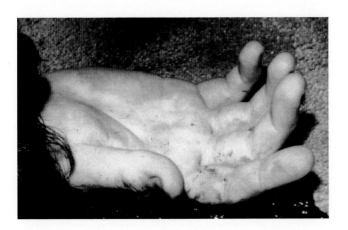

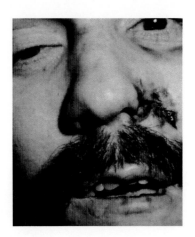

Figure 11-11

Transfer of blood seen between knuckles.

Figure 11-12

Face cleaned for autopsy but showing abrasion between the eyes and bruised lips

It was determined that at the moment of the shot, the glasses fell off. The glasses protected the orbital space during the shot, but a drop settled on the eye afterward. This meant that the victim was on his backwhen shot. Had he been seated upright and fell away from the shot, the settling drops would have fallen on his legs (or chest area), not his face.

Several indications exist that the victim was on his back when he was shot, not propped on his elbow reading. He was wearing the glasses that witnesses indicated he did not need unless reading, and did not wear when he was drunk. According to witnesses below the apartment, an hour elapsed between heard arguments of a fight and the gunshot. Immediately after the shot the body would have had to be repositioned for the blood flows to mark the final resting place as they were found.

The kitchen sink had scattered bloodstains consistent with PAB/Mix-Water. Colors of the individual spatters were from red to salmon pink (Figure 11-17, see page 120). A portion of the counter top was broken off and found on the floor. Several arrangements were tried to determine how the sink stains could result. Experimentation showed a better distribution of even color for the stains resulted

Figure 11-13

Experiment with drying time for glasses. Note the angle and contact necessary to keep blood from the orbital space.

from holding a block of ice in bloody fingers and throwing it at the sink. There was ample indication that a fight preceded the gunshot and blood flowed while the victim was still capable of movement and breathing.

ADJUDICATION AND RESOLUTION

Finding other evidence for corroboration is usually easier when the bloodstain pat-

Figure 11-14

20 minutes of drying is not enough. 45 minutes of drying at 85° F necessary for no flow into the orbital space.

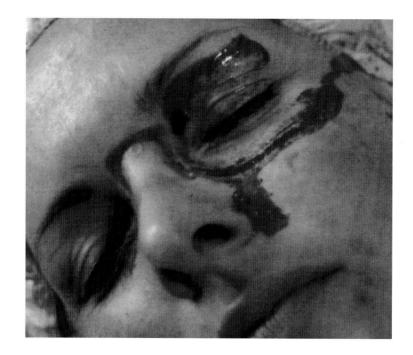

Figure 11-15

Constructed reproducible head pivot device.

tern interpretation has been correctly completed first. If a subjective approach is used, the BPE will simply be fit into a preconceived scenario based on other evidence. If the scenario is wrong, the benefit from a bloodstain pattern analysis is lost. This was the case here. The occurrence of an altercation prior to the gunshot was confirmed with witnesses, and the fact that the victim was moved after the shot supported it. These could be confirmed in at least three ways with other evidence as well as BPE, yet trial strategy relied upon a scenario presented near the beginning of the investigation. Bloodstain pattern information was applied to confirm the earlier assumption, not as investigative leads or independent corroboration.

The victim was alleged to be sitting on the floor listening to records when he was approached and shot. The argument was that the stains on the table were "inspiration" or HVIS. Since the victim couldn't breathe after being shot they had to be HVIS. The defense, however, had discovery of adequate information showing that there was an altercation prior to the gunshot, thus making respiratory distribution possible. No other bloodstain pattern evidence was used. The jury believed there was adequate reason to doubt the prosecution case and thus accepted the accused's claim of self-defense. They voted for acquittal.

WHAT CAN WE LEARN FROM THIS CASE

This case emphasized the importance of the autopsy report in the interpretation of the

Figure 11-16
Gunshot into the right nostril area of styrofoam head. Blood was placed in a depression in the head before the shot. The same experiment was repeated four times without cover over the blood source. Every time blood was projected into the orbital space.

crime. The amount of blood in the lungs should have cautioned any bloodstain pattern analyst to be aware of the possibility of respiratory type patterns as well as those from the gunshot. In the past too many professed experts have ignored autopsies when analyzing cases. This is becoming less common as experts are obtaining better qualifications and experience with the evidence.

Law enforcement experts may at times put excess emphasis on suspect and witness interviews in preference to analysis of the crime scene. That was not the situation here. The witnesses' statements that there was an hour lag between an altercation and the gunshot was thoroughly pursued by detectives. The wrong direction was encouraged by the original scene processors, who claimed that the **string reconstruction** (photographs not available) confirmed an earlier two-dimensional sketch on a crime scene photograph. The string reconstruction located the origin of the gunshot based on the end table and the planter at 5 to 7 inches from the floor. The distance between the victim's face and back of the head was 5½ inches. There was no room for a propped up elbow. The victim was on his back when shot with the possibility of slightly tilting his head up. The glasses came off at the moment of the shot (spot on right eyelid).

Perhaps the most important lesson from this has come long after the case was closed. A new subject area for bloodstain patterns is staging. This is presently not given the time and study that it deserves. Bloody crime scenes may undergo many changes including **contamination** and investigative transfers (IT). These differ and require recognition separate from evidence of delib-

String reconstruction: The use of strings drawn from individual stains to locate the origin in space of an impact event.

Contamination: Changes in evidence that result from actions of bystanders and rescue teams.

Figure 11-17
Blood in a sink resembled
holding ice in bloodied
hands experiment (see
Section V).

erate and premeditated staging. As this and other cases presented here illustrate, staging can be elaborate, simple, or inadvertent efforts to change the context of the crime scene. Recognition of even small changes can be extremely important as investigative leads.

Figure 12-1   View of the artistry at the victim's residence.

CHAPTER 12

CRIME SCENE ARTWORK, STAGED ASSAULT

Case 8

BACKGROUND

At 5 A.M. the newspaper delivery people found a man on his front lawn in a suburban area, the apparent victim of multiple bleeding stab wounds. They notified emergency services who notified law enforcement. The victim was air transported to a large hospital complex emergency department. Paramedics, admitting nurses, and attending physicians all carefully documented his condition from the moment of the helicopter arrival at the scene to discharge after hospital care. He remained lucid, conversational, and apparently completely in control of his words and actions during each recorded evaluation.

The victim related to investigators a story of meeting a man, later identified as a newly released ex-convict, in a rough area bar. The two of them had negotiated an agreement and subsequently went to the victim's home, shared drinks, cocaine, and engaged in gay sex. Sometime in the early morning hours, the victim claimed, the ex-convict attacked him. He described the attack from behind in the laundry room, then the assailant moving in front of him, and his efforts to kick the man. He described being stabbed, falling to the floor, and playing dead. Further descriptions were that the victim crawled to the phone, remembered it wasn't working, and therefore crawled back to the front door. Finally the victim claimed he crawled out the front door and into the yard where he hoped the paper delivery people would find him.

The victim was told, soon after the initial statement, that the police had hired a bloodstain pattern expert to review the crime scene and help reconstruct what happened. Immediately after this disclosure the victim claimed faintness and requested the interview be terminated. Medical reports with time notations showed immediately following the detectives terminating the interview that the patient was lucid, able to answer questions, but in some pain from one injury that nicked his kidney. His complaint of pain was inconsistent with his feeling faint and needing to sleep.

The suspect in the alleged assault was quickly apprehended but had a different story as to events. He admitted he was an ex-convict, and was not gay himself, but had a drug addiction. He claimed the victim approached him in a bar and offered him cocaine in exchange for sexual favors. They left in the victim's vehicle and went to his residence, over 20 miles away. The suspect claimed that they engaged in sexual activity, then used cocaine after which the suspect passed out. When he awoke, he found the victim on the living room floor and the scene all bloodied. He searched the victim's discarded pants pockets for money, drugs, and keys to the victim's vehicle, which he stole in order to leave the scene.

The suspect later tried to discard money with the victim's blood on it and witnesses led police to the discards as well as to the suspect. In rambling testimony the suspect admitted he didn't remember what happened but accepted that he might have attacked the victim as he had done so before to another man while under the influence of drugs.

BLOOD SOURCES IDENTIFICATION

The victim sustained seven cut/stab wounds, none immediately life threatening. Three shallow cuts were delivered to the left forearm positioned almost parallel and at right angles to the axis of the forearm bones. A horizontally aligned cut had been delivered to the left neck, in which the jugular vein was nicked. The jugular vein carries blood from the brain back to the heart to become oxygenated and subsequently return via the arterial system. It is a major venous blood vessel but not pressurized as is the carotid artery, and not as essential to life. No arterial blood vessels were involved. A fifth stab wound was located between ribs on the left side of his abdomen with tracking missing all vital organs. A sixth cut to the scalp was superficial. The only serious injury was a stab wound to the left buttocks in which the knife blade entered the muscle but deflected off the hip bone traveling upward, nicking the left kidney. This led to pain and the presence of blood in the urine, which were the reasons for hospitalization. The victim lost blood but the amount and time interval over which it occurred was not an immediate danger to life. All injuries were to the left side with the exception of a small scalp cut.

BLOODSTAIN PATTERNS IDENTIFIED

Because this wasn't a homicide, and medical records would be slow to reach the detectives, an open read was performed at the scene, i.e., identification and interpretation of bloodstain pattern evidence without any knowledge of blood sources or other evidence of the case. After identification and interpretation, the medical evidence and statements were reviewed, and confirmed what was suggested by the actual crime scene. The scene resembled an elementary school finger painting project (Figure 12-1).

Walking through the residence was first performed to locate where events began. The simplest approach for this was to focus on the amount of blood. Very little blood suggests locations of early events with the first injuries, and copious quantities suggest toward the end when blood is flowing freely from more wounds. This was a good approach in the case because the events moved around. If everything happens in one place, separating events on the basis of quantity of blood may not be possible.

The area with a few bloodstains was the laundry room between the garage and the rest of the residence. A scattered spatter pattern was seen on the side of the washer/dryer (Figure 12-2), set to the right of the walkway facing the entry from the garage. The height and location suggested this as the position in which the victim's stab to the abdomen occurred, if the victim were on his knees beside, facing the shelves behind the machine.

Figure 12-2

Probable impact spatter on the side of the washing machine and dryer.

On the floor with slight direction of travel crosswise to the walkway of the laundry room, were drip cast offs (LVIS, drip trail, passive stains). One spot had been stepped in and was a transfer (contact stain) to a second spot (Figure 12-3). These corresponded to a stain on the bottom of the victim's moccasin slippers (photograph not available).

Along the wall to the left (walking away from the garage exit) were smudges and random spatters, later classified as swing cast offs (Figure 12-4).

In the hall, immediately outside the laundry room, were two slippers. One was at right-angle to the laundry room entry, and the other was in line with the walkway away from the laundry room but positioned with the toe toward the front door entry (Figure 12-5). These were verified as belonging to the victim, and by his statements were dislodged from his feet during the attempt to kick his assailant. Since the victim reported trying to kick his assailant, this action was later tested at a doorway to examine vulnerability to attack at the time. Reconstruction attempts to kick an imaginary assailant while wearing loose fitting slip-on shoes dislodged both shoes as the victim's were found at the scene. However, the exercise suggested that no person was in front of the victim when the shoes were kicked off. Had there been someone there, the shoes would have been blocked from traveling the flight path they had in order to be found as arranged.

Figure 12-3

Drips showing person facing wall then backed up stepping into one of the drips.

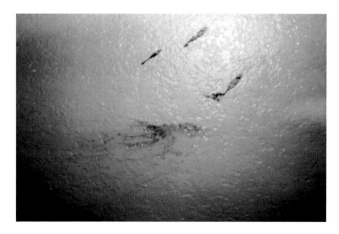

Figure 12-4

Smudges and cast offs on the wall show that the victim's left side is in contact.

Figure 12-5

Shoes positioned as found, claimed to have been lost in attempts to kick assailant.

Figure 12-6

Triplicate direct transfers of the three cuts on the left arm are seen on the wall.

The wall in the connecting hall to the living room had several separated patterns. The most noteworthy were three direct transfers (contact, compression stains, passive), shown in Figure 12-6. These were interesting in that they did not show horizontal smudging (i.e., moving transfer with direction of travel back and forth). To produce the pattern the bloodstained object was touched to the wall, lifted, touched to the wall, and lifted and touched a third time and lifted. The spacing seems very uniform, which would be uncommon but perhaps not impossible occurring once for two men during a struggle. Three such uniform patterns were suspicious. Also, if the victim was being attacked on his left side, how was the attacker reaching him if the side was touching the wall, and why was the exposed, and vulnerable, right side without injury?

A bell-shaped spatter arrangement was seen on the opposite wall, across the short hallway. The individual stains were at slight angles to each other and cascade downward according to gravity in a manner identified with cessation cast offs (passive stains). This pattern may easily be recreated with a paint

brush. Wet the brush thoroughly with blood (or latex paint) then flick it at a surface up to down terminating abruptly by tapping the shaft against the forearm. The blood drops continue traveling downward, spread in a bell-shaped whole arrangement (Figure 12-7). The pattern can result from defense gestures, but such as seen would be those blood drops delivered away from the wall, not deposited on it.

The hall terminated at the juncture of three living areas—living room, family room, and dinette—with a fireplace dividing the space. The front door was reached across a parquet floor off the carpeted surface of the hall and other rooms. From the juncture toward the kitchen and the dinette was a large moving transfer (swipe/wipe, smudge) showing cross wiping all along the line of travel (Figure 12-8). Periodic heavier areas of staining can be seen. The impression from the whole arrangement was that some material objective was used to spread blood, not to remove it.

On the brick of the fireplace hearth is a **simple direct transfer** (contact, compression) characteristic of a hand print. No blood was seen in the palm area of the pattern, but appeared limited to the finger tips of an open fist of the right hand[1] (see enlargement of pattern in Figure 12-9). The victim admitted this was his hand print and claimed that he used the fireplace to attempt to stand upright. However, he earlier claimed to crawl all the way back to the front door.

At the front entry an area of extremely heavy blood staining, as seen in Figure 12-10 (see page 129), was on and around the parquet. Volume blood (pooling) was seen all over this area with various transfers in shag carpet and those resembling terry cloth contact. Closet doors to the left facing the front door showed low-level random, heavy density spatters (dark, almost black) characteristic of splash (splatters) after some degree of clotting (PABS/Clot). Above these is a transfer pattern, which was different from either the shag carpet or terry cloth. It resembled the texture and curls of the victim's hair. The action suggested is head back against door and roll from side to side (Figure 12-11, see page 129). To the left of the closet, on the floor are transfers (voids) where blood has been removed from the volume (pool) stain at a time interval sufficient for loose clot formation (PABS/clot). These were suggestive of fingers drawn through the blood, again after clotting had firmed to the point where the blood does not flow back together (Figure 12-12, see page 130). To the left of the transfers on the carpet at the begin-

Figure 12-7
Bell-shaped whole pattern characteristic of a cessation cast off.

Simple direct transfers:
Touch without movement of a bloodied material to a non-bloodied surface.

---

[1]Wonder, *Blood Dynamics*, 88.

Figure 12-8

The victim described his crawling to the phone and then his crawling to the door.

ning of the hall leading to the walls previously discussed, are spatters arranged in cane shapes, which were flicks, as described in a previous case.

At the threshold to the front door is a simple direct transfer (transfer, contact) shoe print in blood (Figure 12-13). The print is positioned at an odd angle to exiting the door and aligned so that a clear print is seen. No pressure is indicated on the sides or toward the heel or toe. A similar print is found in the hall (Figure 12-14, see page 131). No drips or other transfers are seen leading to or away from the isolated prints. No moving transfers (swipes, wipes, or smudges) are identified at the door threshold.

At the entrance to the laundry room was another hall connecting to the bedrooms and a bathroom. In this bathroom a few spatters were found on the wallpaper and a smudge on the light switch. There was an absence of drip cast offs in both of the halls leading from the laundry room, as well as an absence of blood in and around the faucets.

Figure 12-8

The victim described his crawling to the phone and then his crawling to the door.

OBJECTIVE APPROACH TO INTERPRETATION

The first observation was the arrangement of patterns in the laundry room. Since the examination of the evidence occurred before detectives talked to the victim, they had the information before the victim was aware that his story could be checked. When asked where the attack started he confirmed that it was in the laundry room. The stab to the abdomen appeared recorded on the side of the washer, but would have required the victim to be on his knees with his face too close to the wall for an assailant to be standing in front of him. The position the victim was in when the wound opened was not only impossible for an assailant to reach, but exposed his entire back. No injuries were on the victims' back.

Figure 12-9

Patterns such as this suggest something held in the hand when exposed to blood.

The victim claimed the attack started from behind before his pants were removed, but no tear, cut nor any blood was found on the slacks in the hip area as he was allegedly wearing. The pants were found by the front door with a small amount of blood in the pocket area. The hip wound was the only injury that could have been delivered from behind, but must have been when he was no longer wearing the pants. He alleged the pants came off in the living room later, but did not claim he was stabbed after the pants came off. There was no cut or tear in the hip region of the pants. In fact there was no blood on them except in the pocket area. The suspect admitted reaching into the pocket of the pants for keys and money.

The next pattern in sequence was the drip cast offs. Here the victim had risen and stepped backward so the assailant wasn't behind him. The question follows, where did the assailant go between these events? The victim claims to have kicked his assailant, and again the action is indicated in the position of the slippers, but there was no one in front of him when he kicked.

Next consideration was the cuts on the forearm because they left evidence on the laundry room wall. If the victim is facing the attacker as he's bumping against the laundry room wall, how does the attack reach the arm against the wall? Further, why not attack the exposed right side?

The three parallel cuts on the left forearm are suspect because of their symmetry and the repetitions of transfers to the wall. Again the right side is fully exposed with no injuries. Outside the laundry room the transfers are not consistent with a struggle composite. Composites frequently involve swing and drip cast offs smudged and moving transfers, but here is a lack of the rapid motion type of cast offs seen in defensive gestures. The actions are smoother, appearing measured and separated after apparent forethought (i.e., touch, lift, reposition, touch, lift, reposition, and a final touch).

Figure 12-10

Transfer patterns from the long-fiber shag carpet and the short-fiber terry cloth.

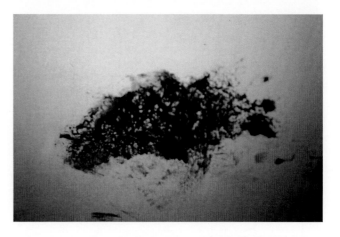

Figure 12-11

Hair transfer with head rolled side to side while leaning against closet.

Since cast offs were definitely identified, we'd expect drip cast offs on the carpet near the wall patterns. None were seen next to the transfers. Drips were seen on the opposite side of the hall in cane or flick arrangements but associated with the bell-shaped wall pattern, not the transfers.

At this point we can account for four injuries, the left abdomen and the three cuts on the left arm. These were positioned so that they match the heavy areas of the wide stain on the family room carpet. The direction of travel for this whole pattern was across the direction of alleged crawl. The blood seems to have been wiped across the stain all along its path. The concentration was not distributed from areas of heavy staining. The pattern looks like blood was spread to increase the surface covered, not to clean it up, yet the crime scene photographs were taken before the victim or anyone had access to the scene.

Figure 12-12

Volume blood that has initiated clotting so finger marks do not flow back.

Although the sequencing of the cut to the scalp and cut to the jugular can't be confirmed, the location of the latter can be suggested. The stains in the bathroom were at the height and position that would occur if the victim were standing in front of the mirror to see how the cuts were made. Again if this happened, there was no logical position for the assailant to have delivered the cuts. The logical interpretation is that the victim watched the mirror as he cut himself.

Perhaps the most important pattern of all is the hand print on the fireplace hearth bricks. The significance of this was missed by several investigators, including the author. A knife was in the kitchen sink but discounted as the weapon because the victim said it wasn't the one used by the suspect. The hand print was matched to the victim, who stated he used the brick hearth to pull himself up from the floor, yet there is no evidence of blood on the palm part of the print. The blood was on the tips of the fingers as it would occur if the victim were holding the knife himself.

Figure 12-13

The position of the shoe print, linked in class characteristics to the suspect, seems awkward if it was the result of stepping out the door. Exit used was garage, not front door.

The victim claimed to crawl to the phone, then crawl back to the door. No indications were found to show he crawled from the phone. It was more reasonable to accept that he stood up by the fireplace and then walked to the door.

A large shag throw rug was found against the front door, so the question is asked, how did the victim get outside? His story was that the assailant left via the garage in his stolen vehicle. It's possible that staging the scene got a little confused and that the rug against the door was to support the story of the assailant leaving via the garage after the crime. The victim may have become concerned with too much bleeding and left by the garage door so as to not disturb an earlier staging. Outside the front door there is a gap from the very

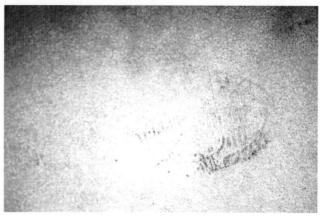

bloody entry area and the first stoop. Blood drip cast offs begin again on the sidewalk (Figure 12-15) and drops were falling frequently enough that one would have expected to see them from the door onward.

Unfortunately, the victim's statements were considered of more importance than the bloodstain patterns. The shoe prints were of the suspect's shoes. The lack of stains leading to and from them was ignored, as was the lack of a blood source for the shoeprint found on the hall carpet.

Figure 12-14

No blood source in the hall where this is found and no other prints before or after. Other investigators noted that there was no pressure, thus not indicating shoe was worn.

ADJUDICATION AND RESOLUTION

The suspect pled guilty to the assault and no charges were brought against the victim. The ex-convict had difficulty adjusting to life outside of prison, so readily agreed to a guilty plea and reincarceration. The police were happy because their good work was rewarded with a "bad guy" back in prison, and the prosecution felt the same. The public defenders were too busy to care and felt good because the client was satisfied with his sentence. The public probably benefitted because the accused was headed for more crimes to feed his drug habit, and he admitted that he'd gotten by with a violent crime in the past. The only person unhappy with the outcome was the bloodstain pattern analyst.

Figure 12-15

Blood drip cast offs begin on the sidewalk.

WHAT WE CAN LEARN FROM THIS CASE

The biggest lesson the author learned is that a true objective analysis is not always welcomed. Such can happen with either side, prosecution or defense. It's idealistic to feel that experts be totally impartial when their very livelihood may depend upon seeing any given case from a specific view. It has come to the author's attention many times over the years since this case that if attorneys, on either side, do not get the testimony they want, they will often hire someone else who is willing to say what they are told. If being an independent consult is the only way to put food on the table, there can be no doubt that a subjective approach must be resorted to. The only possible answer is to pass laws that would require experts to answer to and be hired directly from the courts.

A good lesson endorsed here also is that victims do not always tell the truth. It's always good practice to view the evidence before talking to witnesses, suspects, and victims. Inconsistencies should be examined not ignored. Most law enforcement officers are highly suspicious people but such discrepancies as an ex-convict versus someone with no previous contact with the law can prejudice an investigation.

Figure 13-1   Very steep roadside berm with arterial gushes from decapitation by gunshot.

CHAPTER 13

CURIOSITY CAUGHT THE MURDERER

Case 9

BACKGROUND

The victim was a female coworker of the suspect. She was abducted from work, taken to an isolated place along a rural road where she was raped in her own vehicle, then taken out of the car and shot. Her body was carried or thrown (according to the suspect) over a barbed wire fence and left near a hazardous, rocky waterfall. The bloodstain patterns were noticed on the roadway by utility service employees, and law enforcement was called. While the crime scene was being processed the next day, the suspect drove past the area in the victim's stolen vehicle. Officers recognized the car, and after a short pursuit arrested the suspect.

Immediately after completion of processing the scene the fire department was called to wash away all bloodstain patterns, including a large V-shape visible from the air.[1] Processing therefore was done completely with photographs. The scene of the gunshot was on a steep roadside berm (Figure 13-1).

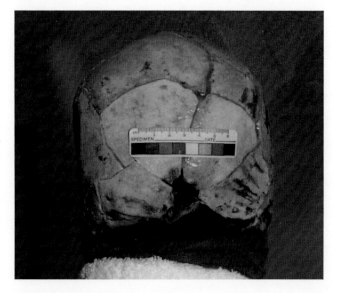

Figure 13-2
The reconstruction of the skull to identify gunshot to back of the head.

BLOOD SOURCES

Despite injury from the rape, only one event appeared to provide a blood source. The victim had been shot in the head, which resulted in decapitation. Both interior carotid arteries were severed with removal of the top of her skull. The victim was wearing a full mask ski cap at the moment of the shot, which contained much of the initial impact. The forensic pathologist involved with the case did a remarkable job of reconstructing the skull of the victim showing the bullet entrance (Figure 13-2).

---

[1]Wonder, *Blood Dynamics*, 79.

BLOODSTAIN PATTERNS IDENTIFIED

The obvious pattern seen in photographs of the scene was the arterial damage (spurt, gush, streaming) (Figure 13-3). The whole pattern extended 14 feet and was described as a V. Upon closer examination of the tip of the V, the whole pattern shape was better described as a W, with the first and last lines greatly extended. Between the inside lines of the W was a slightly undulating row of large, equal-sized ovals in a parallel arrangement (Figure 13-4). A close-up of this (Figure 13-5) can be compared to a close-up view from Figure 1-1 of a row of ovals seen in the 17,000-year-old cave painting at Lascaux, France.

Figure 13-3

The V described in reports was actually a W with longer side branches.

Figure 13-4

Close-up of parallel ovals between branches of the W.

Smudges were found in the car, and small spots of blood that could be spatters, dried flakes, or secondary spatters were scattered along the chrome running board on the passenger side. By the time the items were examined it was not possible to completely classify these stains with any degree of certainty. They were grouped as secondary to those events that defined the crime.

Transfers were found on the passenger seat. When access to the suspect's clothes was finally permitted, no bloodstains were found on the seat of his pants. No blood was noted from the victim's buttock at the time she would have been in the car.

A full-face ski mask was placed over the victim's head (Figure 13-6). Directly below the face opening, chin/neck area of the mask, was a single row of arterial spurts (photograph not available). An identifiable hole was seen at about the temple region on the left side of the mask. Turning the mask inside out exposed considerable brain and small bone fragment debris adhering from the decapitation. Two holes were found on the inside of the mask with one, noted previously as the bullet entrance, penetrating the entire weave, and the other hole penetrating only the innermost weave. This second incomplete hole, located roughly three inches up and slightly to the right, was surrounded by black residue. Evidence that the partial hole existed when processed by the crime lab is seen in the lab photograph as a slight halo in the shadow (Figure 13-7), which confirmed the presence of the partial hole at the time the mask was photographed at the crime lab. Figure 13-8 shows the reverse of the mask many months later when debris was dried.

OBJECTIVE APPROACH TO INTERPRETATION

Random spatters are seen on the victim's hands as they are tied behind her back. These are important when compared to spatters on hands that are connected to suicide by a gunshot to the head (Figure 13-9). The presence or absence of stains alone should not lead to conclusions, nor assumptions based on spatter size alone. It is equally important to note directions of travel and arrangement of spatters within a pattern. These stains resulted from fallout of arterial damage stains, termed **arterial rain**, following the gunshot to the head. They show that her hands were tied behind her back at the moment the carotid arteries were breached, which in this case was the moment following the shot.

The central pattern to the crime scene was the V/W arrangement on the road berm. Investigators initially focused on locating the origin of the gunshot. All the bloodstains identified were from the carotid arteries. The row of large ovals was a transition from one position of the head to another, i.e., the victim's body was pivoted before collapsing to the ground.

The pants of the suspect show a rip along the ankle area and his own blood was identified as dampening the edges. Taking into consideration the foot, shoe, and ankle height, the tear is at the same level as the bottom row of barbed wire at the crime scene. The accused acknowledged that he and the victim were at the fence together. This tear with his blood positions them at a time when the accused could have sustained an injury from the barbed wire. This coupled with his demonstrations of temper during pretrial appearances could explain the gunshot being fired where it was.

The holes on the inside of the ski mask showed the angle of the shot when the gun was fired. In fact they also explain why the bullet shattered upon entrance, and subsequently fragmented so that complete decapitation resulted. It is stated in reference material[2] that bullets may not fragment extensively upon entrance but do so after hitting a second surface. In this situation the bullet was split by shaving at the edge of the skull cap. After shaving, the bullet (missile, projectile), fragmented and pieces exited around the skull cap. The results of the shot initially were contained within the ski mask. After the shot, the mask was thrown across the fence with the contents distributed in an area some distance from the fence. It was misinterpreted by investigators that the gunshot itself had distributed all the fragments. The distances involved were unrealistic for a gunshot to project the amount and size of the brain fragments.

Arterial rain: A horizontal arterial pattern resulting from fountain fallout.

Figure 13-5

Close-up of row of ovals from 17,000-year-old cave painting in Figure 1-1.

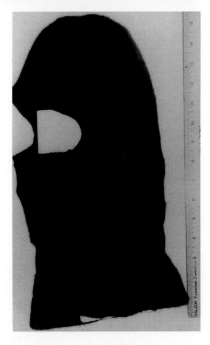

Figure 13-6

A full-face ski mask worn by the victim at the moment of the shot. Face opening was to the side.

Figure 13-7

Lab photo shows a shaded area that corresponded to an exit partial hole.

Figure 13-8

Inside of the full ski mask worn by the victim at the moment of the shot.

Figure 13-9

Spatters on the victim's hands are from the spurting carotid arteries not GDIS.

The investigating officers originally alleged that the victim was shot execution style. This was based upon misapplication of information from a Bloodstain Pattern workshop final exam question. On the exam a crude sketch of a pillow, extruded brain, and blood volume with bilateral arterial spurts was used with a list for multiple choices of the wording for the observation. Detectives remembered the sketch when seeing the present crime scene. They applied a pattern match approach to the case and concluded the answer to the exam question fit this case since the crude picture resembled the crime scene evidence. The error occurred because spatter patterns are not pattern match evidence. The sketch question was designed to detect subjective labeling of evidence, not to be used as a memorized pattern. To use the picture to draw conclusions in a case was the opposite of the purpose in the exam. The case did have two arterial damage patterns but the scale was considerably different, and the two sides of the V were actually both carotids in two positions, not one on each side of the theoretical head as in the sketch. The arrangement of the ovals within the patterns was different. Interpretation of each pattern group must be within the context of the crime, not on the basis of one looking like another. In later classes the exam question was deleted because it had failed to achieve its objective.

In this case the location of both carotid arteries gushing in two different directions with a pivot between them showed that the victim was standing

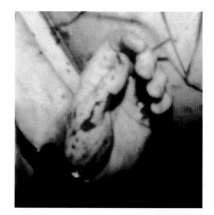

[2]DiMaio, *Gunshot Wounds*, 118.

upright at the moment of the shot and
pivoted through approximately 130
degrees or so during fall. She could not
have done this on her own. This also
led to the suggestion that the gushing
arteries may have hit the back of a second person present during the crime
events. When her body bent forward
the arterial gush could have projected
blood onto the hip area of the second
person, which was then transferred to
the car's passenger seat. Smudges were
found on the car passenger seat from
that transfer or from smearing spatters
later. No blood was found on the suspect's pants hip area. If the spatters had
been on the seat originally, someone else
smudged them. The spread and location suggested the stains were deposited

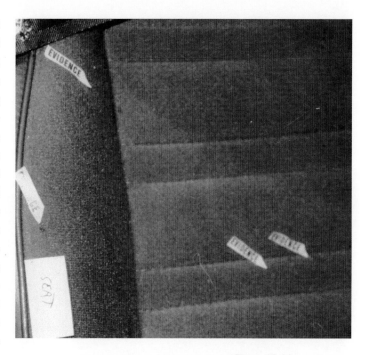

Figure 13-10
Smudges on the passenger side car seat.

not smeared spatters (Figure 13-10). Either way, a second person was present
during the events at the crime scene and at some point after bloodshed, sat in
the passenger seat.

ADJUDICATION AND RESOLUTION

The author was given two stories regarding resolution, one was that the case
pled and another was that it went to trial and the suspect was convicted.

THINGS TO BE LEARNED FROM THIS CASE

The first point to be made is that although death was by gunshot, no evidence of
gunshot distributed impact spatter (HIVS) was found, nor would it be expected
when the victim was wearing a thick-knit ski mask at the moment of the shot. The
original CSI on the scene found HVIS in the grass of the berm because he was
specifically told to find it. The sequences of events and information regarding
the shooter were available with bloodstain pattern evidence, although it did not
include evidence of the gunshot itself. Looking for one classification of pattern,
HVIS, can prejudice an investigation causing investigators to miss other important investigative leads. In this case the presence of a witness or second participant
to the crime was more important than finding where the gunshot occurred.

Figure 14-1   View of crime scene in the out-building of a commercial field.

CHAPTER 14

INFORMANT EXECUTION

Case 10

BACKGROUND

An illegal immigrant was running a drug trade using migratory farm workers as mules. One of the mules came to the leader's notice as a suspected informant to law enforcement. The drug dealer, the suspected informant, and another man walked into a field urinal house. The suspected informant was pushed to the ground and shot in the head with a .22 handgun. The leader and his other employee then left the area. The drug dealer left the country.

Later, the third man was arrested on another warrant. To bargain down his charges he supplied information relative to the execution of the suspected informant. Law enforcement wanted confirmation of the statement since it was believed that lies would be told simply to obtain a more favorable plea agreement. A bloodstain pattern analyst was taken to the scene of a cold case.

BLOOD SOURCE IDENTIFICATION

No details were provided initially except that the visit to the scene was of a homicide by small caliber gunshot wound to the head. The autopsy and photographs of the scene were supplied for confirmation later. No photos were provided until after the scene was processed for investigative leads information. The approach would be as a blank read for bloodstain pattern evidence.

Given the cause and manner of death, we would be aware of the possibility of patterns consistent with gunshot distributed impact spatter, but not look for them exclusively. Subsequent occurring patterns such as volume and drip cast offs would provide endpoints to help backtrack to the general location of a homicide. If no body is present, and nothing is known regarding the crime, the best place to start is with a volume (pool) bloodstain. This is actually preferable to beginning to process a scene based upon witness statements, because the latter assumes the witness is truthful and has knowledge of the crime.

In homicide investigations, generally someone is lying, and it may be a witness. It is better to approach the scene with disbelief in all statements. Lies uncovered at the first interview have a better chance of resulting in confession than lies uncovered after a signed statement.

Figure 14-2

Metal sheet where three areas of convergence were identified for GDIS.

Figure 14-3

Enlarged pie-wedge-shaped spatter arrangement identified as GDIS.

A cement block building, used as a field urinal, had been constructed of two rooms, with one covered in a cement floor and the other with hard packed dirt. There were holes for toilets, but no commodes present. The dirt floor area extended beyond the toilet area with an old nonfunctioning water heater (Figure 14-1). Along the wall was a sheet of metal that had served as flashing behind the heater. A hole in the sheet metal was not associated with this crime (note dowel protruding). On the dirt floor was a volume (pool) stain and drip cast off (drip trail, passive stains) in a scattered arrangement not showing exit of the blood source for the drips. Since these were in relatively close position to the steel sheet, it was examined in detail using a page (2×) magnifier.

Three separate impact spatter (HVIS) patterns were identified on the steel sheet. All three showed all the characteristics of an impact pattern consistent with gunshot distributed bloodspatter. The whole pattern shapes are a pie-wedge with individual stains at angles to each other, and at angles to the whole pattern. The sizes were all relatively small, visible to the unaided eye, but did not show up well on the photograph of the metal sheet (Figure 14-2). Enlarged views show the pie wedges located close to each other but not sharing a common origin (Figure 14-3).

OBJECTIVE APPROACH TO INTERPRETATION

The fact that there were three wedges on one surface was good to limit positions for the victim at the exact moment of the gunshot. The autopsy information provided the location of entrance and exit wounds as well as tracking of the bullet between each.

Two of the wedges were in close proximity with what could be the same apparent area of convergence, and the remaining wedge was separated by a few inches from the two. The two wedges suggested bone (such as cheek or zygomatic ridge) located near the GDIS/entrance (HVIS, back spatters, blow back spatters) causing the split into two patterns from the area of distribution. A fragmenting bullet could also result in multiple GDIS/exit (HVIS, forward spatter) resulting in two to three patterns. The separation into two and one wedges suggests entrance and exit wounds, although all three convergences were located on the single plane of the metal sheet.

A string reconstruction was performed and general origins for each of the impacts located. These were correlated with the volume and drip cast offs on the floor. The author then positioned herself for entrance, exit (based on autopsy information supplied after the reconstruction) and spatter patterns, as the victim had to have been at the exact moment of the shot to the head (photograph not available).

Figure 14-4
Position of the body as found.

ADJUDICATION AND RESOLUTION

The position of the victim at the moment of the gunshot identified from bloodstain pattern evidence was valuable corroboration for the witness statements of what happened in the field house. The body was found face down (Figure 14-4). When the body collapsed after the shot, the position shifted with the victim's arms crossed under the body and was revealed only after the body was rolled over by coroner's officers. This same position was duplicated by the author acting out the position from the string reconstruction and pantomiming collapse from execution.

The only people who knew the arms were crossed under the body were the lead detective, the identification technician, and the coroner's deputy (Figure 14-5). No one else was told. The position the victim was in to achieve the arms crossed confirmed the statement of the witness regarding the relative positions of the three men at the moment of the shot.

Figure 14-5

After processing the scene the body was rolled over and it was found that the arms crossed underneath. This was predicted after a string reconstruction with knowledge of entrance and exit wounds.

WHAT WE CAN LEARN FROM THIS CASE

We learn that bloodstain pattern evidence can provide investigative leads information to use with eyewitness interviews. Too often the applications are limited to application after all other evidence is evaluated. This may even lead to bias as the bloodstain patterns are then "shoe-horned" to fit within the context of other information. If that information is correct, the bloodstains add nothing. If, on the other hand, other evidence and witness's statements are false, unreliable, or impeachable, the bloodstain pattern evidence is lost, or even detrimental to the case. When a valuable form of evidence is available, and free, it's a waste and inefficient to not use it correctly.

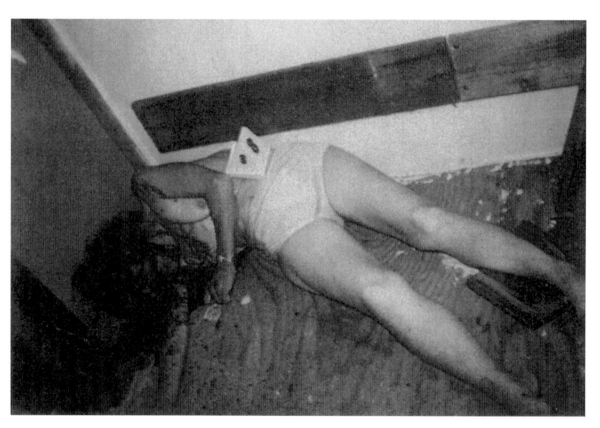

Figure 15-1   The body as found. Impossible shot to top of the head. Visible in photographs at the law enforcement evidence storage was an arterial spurt on the wall to the right (as viewed) of the body.

CHAPTER 15

LACK OF THE CRIME LAB INVOLVEMENT

Case 11

BACKGROUND

Police responded to a 911 call at a mobile home park. The call came from the part-time roommate of the victim of a homicide. The roommate, herself, was injured from a gunshot wound to the chin. She claimed a third woman, the shooter, shot the victim inside an unfinished mobile home bathroom, then left the scene. As the shooter was walking across a field to her car, behind the mobile home park, the victim's roommate was struggling with her, searching for drugs and money on her person. The gun, which had been used to shoot the victim, discharged, injuring the roommate.

Law enforcement officers responded to the mobile home park and found the victim in the bathroom. The roommate stated that she had waited for officers outside after calling 911. No volume stains or blood into blood was seen near the deck chair where the roommate supposedly waited.

The shooter had driven away after the confrontation, and the injured roommate returned to the crime scene and called 911. It was later disclosed that one hour elapsed between the time of the first shootings and the 911 call. The time was never accounted for. When the accused was later apprehended, she readily admitted firing a certain number of shots. She claimed to have great fear of the victim and shot her to prevent further threats on her own life. The shooter (now the accused) had traveled to the mobile home park for the expressed purpose of confronting the victim and had taken a handgun allegedly for protection, but did not deny thoughts that she would have to shoot the victim.

The gun was never recovered. The accused related that she saw the victim in one corner of the bathroom, and when the victim moved rapidly, she shot. The victim did not seem impeded by the first shot so the suspect fired several more times in rapid succession. The last view of the victim the suspect remembered was of her squatting at the farthest wall near a louvered cupboard and to the side of a commode. Officers found the victim laying against an inside wall in a space where a bathtub was to be installed (Figure 15-1).

Figure 15-2

Autopsy showing perforating wound to the left temple area of the victim.

The roommate was a witness but denied seeing the actual shooting. She claimed to hear several shots, but did not know how many shots were fired. The suspect stated the number of shots she thought she fired, but was unsure all of them were toward the victim. There were shell casings on the floor but officers noted in their report that they were not sure they had collected all of them. No crime lab was involved mainly because the officers gained a confession.

BLOOD SOURCE IDENTIFICATION

Photographs were the predominant source of information for this case. Articles of clothing worn by the roommate were viewed at the law enforcement storage facility later. Although several rounds apparently were fired from a handgun, only three areas of bleeding could be located from the photographs and study of the autopsy report. Two areas were associated with the victim and one with the witness. The first area noted on the victim was to the left temple (Figure 15-2).

This was illustrated at autopsy with a probe showing complete perforation of the scalp and bone of the temple. The temporal artery is close to the surface in this area and could be expected to have been breached by any bullet passing through that part of the face. No arterial injury was mentioned on the autopsy report when it was made available. Under cross-examination at trial later the pathologist identified the temporal artery as having been breached.

A second area of injury on the victim was the top of the head. Here, tracking of the bullet was described as up to down entering the brain stem. No direct conclusion was provided regarding relative mortality of either wound. Previous case experience, however, suggests that a person may survive injury to the temple even with exposure of the orbital space (see Case 2 in Chapter 6). A shot through a brain such as this, even though life may exist for moments afterward, would be expected to result in immediate loss of voluntary function followed by death without regaining voluntary mobility (Figure 15-3).

The witness was photographed in a hospital bed apparently after treatment and admission. The injury was to the lower chin involving the mouth and accompanied with a good quantity of bleeding. The injury was not considered by medical staff to be life threatening. Investigators and photographs confirmed that no volume blood (pool) was seen where the witness claimed to sit while waiting for police.

Figure 15-3

The injury that would be expected to lead to immediate loss of voluntary functions.

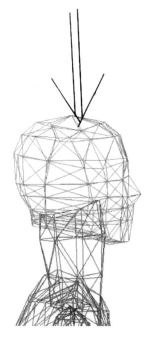

BLOODSTAIN PATTERNS IDENTIFIED

Two volume stains were seen in the unfinished bathroom, with one suggesting that the blood source was bleeding rapidly or with some degree of force.

One volume was located at the furthest wall to the bathroom entrance and the other was associated with the final resting position of the body to the left of the entrance at a position where the tub normally would be installed. The latter also involved an arterial damage pattern on the wall next to the body (no photograph available, views seen at the evidence storage facility).

The most important pattern in this case was arterial rain (arterial spray). The identification is from a random arrangement of relatively large round bloodstains that lack directions of travel. These were scattered over the bathroom floor between two volume stains (pools), and even on the person of the victim (Figure 15-4).

Figure 15-4

Arterial rain, not smudges, not blocked over bathroom floor between positions.

Because crime scenes are often extremely dynamic places during an assault, drip cast offs (drip trails, passive stains) may be scattered around the primary incident such as this but will show some noticeable directionality. Even with perfectly round stains, one edge will often have irregularities or a few spines. Arterial rain, however, results when a breached pressurized arterial vessel projects blood upward. When the drops reverse and fall by gravity, direction of travel is downward at roughly 90 degrees without indications of directionality. The size of area covered by these large round drops indicate the artery projecting the fountain was under high pressure. The temporal (interior carotid) supplying the brain with blood under the conditions of the confrontation could be expected to involve considerable pressure and blood volume, and provide the amount of blood seen in the bathroom.

On the left athletic shoe of the witness is an arrangement of spatters with direction of travel from toe toward arch (Figure 15-5). When the shoe was reviewed later, it was determined that the origin of these spatters would require a horizontal distance of over two feet out of plumb. If the chin wound were responsible, the witness was extending her body far enough outward to have fallen on her face. No transfers, smears, or smudges were found to suggest this and no dirt in the wound indicated a fall in the field after the confrontation.

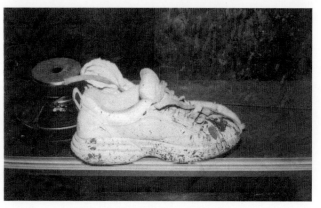

Figure 15-5

Athletic shoe worn by witness at the time of the incident.

Since no volume (pool) stains were seen on the cement pad, it can be concluded that the spatters on the shoe were not blood into blood. The source of the stains was not identified.

OBJECTIVE APPROACH TO INTERPRETATION

A GDIS is seen on the wall area above the victim's body. This confirms that the fatal shot was fired while the victim was suspended above the area where her body was found. It also confirms that the victim was not dead nor unconscious following the shot to the temple because she had not yet fallen to the floor. Two large volumes of blood were seen on the floor by the cabinet area in photographs viewed at the law enforcement agency later, which showed a lapse of time for the victim in at least one other area of the bathroom.

We have three indications of arterial damage, although no mention of such was included in the autopsy report:

- Area of damage in the temple region of the head
- Copious bleeding seen in two pools and random very large, bright red spots
- Identification of arterial rain

The injured area of the temple was so large that a bullet causing the damage would be expected to also breach the temporal artery. This leads to two interpretations. Copious bleeding is often found around head injuries, but this also involved rapid and forceful projection of blood as seen around one of the volume stains. Arterial injury, however, expected for bloodstain patterns seen, must be verified with a medical professional. Once it was confirmed that arterial injury occurred and was in the temporal region, the identification of arterial rain may be made.

The rain was scattered between the volume (pool) stain by the unfinished cabinets and the final position of the body. No smearing or smudging had altered the spots, nor were there tracks or absence areas seen within the bathroom. It was further noted that the position of the victim's body was such that the tracking of the bullets creating the two areas of injury could not have been delivered in the position the victim was found. This leads to three conclusions:

- The shot to the temple came first (it alone could occur without immediate unconsciousness).
- The victim moved from the cabinet area to the tub area herself (no one was standing in the bathroom to block, smear, smudge, or transfer blood from the arterial rain).
- The victim was then shot to the top of her head before falling to the floor as found.

After determining these, we look at statements to determine accuracy and truthfulness. The accused admitted shooting the victim; however, she claimed to have left when the victim was squatting by the unfinished cabinets in the neighborhood of one of the volumes. This was confirmed by the volume (pool) because the victim had to stay in that position long enough for blood to accumulate. Both the witness and the accused claimed the whole incident of initial shooting was over within a few minutes. The overturned fan also confirmed that the victim moved from the cupboard area to the tub area on her own, stumbling as she went.

The witness, however, claims not to have seen anything. Her sneaker shows that she was in the proximity to blood being distributed or projected toward the instep of the left foot. If she was not present nor in alignment when the shots were fired, the only other time she could have received these stains was after being shot herself. The pattern on the shoe, however, suggests that it was not the witness's blood but rather the victim's blood projected either by arterial damage toward the witness or a gunshot delivered by the witness. The gunshot to the top of the head had to occur before the victim fell as found. This does not mean that the suspect did not kill the victim. Blood loss from arterial damage alone could have led to death. The possibility, however, existed for a subsequent gunshot to the head to have been fired after the suspect left the scene. The very least that happened was that the witness did not call for emergency help immediately and may not have been truthful regarding her involvement.

ADJUDICATION AND RESOLUTION

The defense attorney brought out in testimony that arterial damage did occur. In testimony the pathologist stated the exact same number of gunshots that the suspect told the police she had fired. Nothing in the evidence collected nor injuries listed in the autopsy supported the exact number of bullets fired. Pathologists make strong witnesses, though, and this was not challenged during trial.

From the bloodstain pattern evidence it was suggested that the accused shot the victim and left. One hour elapsed between the confrontation and the witness's 911 call to police. During this time a number of events could have occurred. No weapons were available because the weapon used was taken from the scene. No search for alternative weapons was made. No involvement of an excellent crime lab was made due to the suspect issuing an immediate confession. The accused accepted a negotiated plea of second degree murder in part due to alleged remorse of her actions. It was also later learned that the witness inherited all the victim's estate and belongings.

WHAT WE CAN LEARN FROM THIS CASE

This primarily reinforces the knowledge that witnesses can and do often lie, forget, mistake facts, and/or provide incorrect information. Accused people may get confused, be in shock, and lie despite not being guilty of the crime, and remember events differently without an intention to lie. Bloodstain pattern evidence can be applied immediately to the crime scene and used to refute or confirm veracity from the first interview on.

If an agency has access to a good crime lab, which appreciates bloodstain pattern evidence, it seems unfortunate that they would not use it in all death investigation situations. It is recognized that time and funds are a major issue in attempting to limit casework to specific homicides. Practice, however, makes efficiency an economical compromise.

Figure 16-1   View of the final position of the victim. Witnesses watching the confrontation claimed this was the position in which the gunshot was fired.

CHAPTER 16

ON-DUTY OFFICER-INVOLVED SHOOTING

Case 12

BACKGROUND

An officer on routine graveyard shift patrol noticed a driver weaving in and around traffic, suggestive of intoxication. He signaled the driver to stop and was led on a short chase that terminated in an apartment complex parking lot. As the officer approached the car, a second man drove up and displayed an attitude to interfere with sworn duties. The officer called for back-up and then became involved in a struggle with the second man. A K-9 unit responded as back-up, with the dog sent to control the struggling male. At some point in the struggle between officer, interfering man, and dog, the officer's sidearm was discharged.

The officer claimed that the suspect had hold of the firearm and must have pulled the trigger. Onlookers, a crowd of people who were not usually sympathetic to law enforcement, claimed that the officer shot the man execution style while he laid on the ground. The officer was put on leave and the case sent to Internal Affairs (IA). The homicide unit of this agency was asked to supplement the IA investigation with bloodstain pattern analysis. No case photographs were released, yet this is an important situation and we can learn from it.

BLOOD SOURCE IDENTIFICATION

Review was from black and white photographs and with cooperation of an excellent forensic pathologist. The primary blood source was a gunshot to the head from a .38 hollow point, semijacketed bullet. The victim's head x-ray showed tracking of the bullet with fragments, bone and bullet, lining the whole path (photograph not available at this time). There was no exit. The head wound provided rapid bleeding but no arterial blood vessels were involved.

The dog left some teeth marks and abrasions on the victim. The forearms had slight abrasions consistent with dog claw marks, but no significant sources of bleeding were found, other than the gunshot wound to the head.

BLOODSTAIN PATTERNS IDENTIFIED

GDIS (HVIS, blow back spatters, back spatters) were seen on the victim's right forearm, recognized initially on the basis of their apparent point-sized convergence and shape as streaks. Classification was confirmed as impact because no arterial damage occurred and the arrangement of the spatters was at angles to each other with an area of convergence not indicative of cast offs. No volume stain (pool) was seen that could serve as a blood source for splash (splatters) or blood into blood, and no respiratory blood source was found at autopsy.

On the victim's back and on both forearms and upper arms were seen simple direct transfers (contact, compression stains). These appear to be transfers in blood of the dog paw prints. The pathologist specifically noted that there were no skin breaks under the bloodstained prints on the victim's back. Some superficial abrasions were seen under the forearm paw prints. Correlating injuries with bloodstain patterns was rewarding.

OBJECTIVE APPROACH TO INTERPRETATION

The main question posed by investigators, for the bloodstain pattern analysis, was "where was the victim at the exact moment of the firearm discharge?" The eyewitnesses all agreed it was an execution and the victim was on his back. The two law enforcement officers claimed the victim was sitting upright struggling with the officer. The autopsy, however, agreed with the bloodstain pattern evidence, which disagreed with both the law enforcement officers and eyewitnesses. The location of soot within the entrance wound suggested a close, but not contact shot, i.e., victim was neither upright nor on his back.

The spatters on the forearm could be drawn to close to a point, which helps confirm them as probably originating from the entry of a bullet that exposed a blood source. These could then be drawn to coincide with the entry wound in the victim. A reconstruction shows the relative position of hand and head at the moment of the shot (Figure 16-2).

Figure 16-2

Artist model used to recreate the arm and hand position at the moment of the shot.

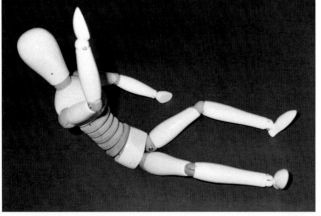

We next look at the evidence from the transfers. Transfer bloodstain pattern evidence may provide two types of information: identity from a pattern match of the transferred imprints, and sequence of actions that have resulted in a series of transfers. In this case we have the dog behind the victim, but also in front of the victim with the paw prints delivered to the forearms but the forearms then transferring blood to the upper arms.

The reason we know this is that there is no blood source on the back, but there was one in front on the forearms. This suggests the sequence of the dog in front, then moving to the back, and the victim trying to push the dog off thus transferring blood from forearm to upper arms.

If the arms were folded in an attempt to protect the head and push away the dog at the moment of the shot, the spatter patterns from the gunshot would have been blocked. On the other hand if the victim were on the ground, even with outstretched arms, the spatters would not have been recorded in the position found. The only position that explains both the sequence of transfers of dog paw prints and the orientation of the GDIS on the victim's arm is that the victim had been fighting off the dog, but at the moment of the shot he was just beginning to fall backward, i.e., in between the two positions noted by eyewitnesses and law enforcement officers (Figures 16-3 and 16-4). If he reached out to grab anything near to stop his fall backward, his arm would be in the correct position to acquire the streaks seen from the gunshot. The officer stated he felt the suspect grab his clothing and thus got a firmer grip on his firearm. The firmer grip could have occurred at the moment the suspect's hold was lost, and the weapon discharged at the beginning of the fall backward.

The eyewitnesses saw the position immediately following the shot, whereas the law enforcement officers remembered the position at the moment just before the shot. Neither group felt they were lying. The soot in the entrance wound agreed with the bloodstain pattern evidence that the victim was inches away from the gun at the moment it discharged. Crime lab evidence put the distance greater than either the bloodstain patterns or the autopsy. Again, stationary shots were fired to locate the position of a dynamic act. (Note the prior example of this in Chapter 8.)

Figure 16-3

Sketch submitted with the report showing dog and victim prior to the shot.

Figure 16-4

Suggested position of victim at the moment of the shot.

ADJUDICATION AND RESOLUTION

This case occurred during a time when bloodstain pattern evidence was just beginning to gain notice. The political manager ignored both the analysis and the pathologist, putting most weight on the residue study and eyewitnesses. The parents of the victim were awarded a sizeable wrongful death payout.

WHAT WE CAN LEARN FROM THIS CASE

Bloodstain pattern evidence is essential to officer-involved shootings because it can provide a nonbiased review of what happened. It is also essential that crime labs be aware of the dynamics involved when interpreting residue analysis from stationary targets. One answer is also to do rotating target shots, although the technique would be new. The actual speed of movement would not be essential as long as the information was submitted as evidence of the differences between a stationary shoot and a moving target shoot. Photographs of the gun barrel are also useful, but one must be aware of angles and more important rapid changes in angle between entry of the bullet and atomizing by the muzzle blast that follows. Research into methodology and techniques of this area would benefit violent crime investigation.

We also learned that eyewitness statements may not fit what actually happened, even though the witnesses may feel they are being truthful. The time of the situation was very late night or early morning with poor lighting. People on the upper levels of the apartment building came out to see the confrontation. Their intentions were not to become involved, thus attention was not focused.

Immediately following the gunshot, the witnesses suddenly were very focused. That was the point they felt clear about, i.e., after the victim was down. Eyewitness statements are not as reliable as bloodstain pattern evidence applied as investigative leads information.

A final point with this or any other officer-involved shooting is that IA officers should have training in bloodstain pattern analysis. Officers being investigated can learn from their mistakes as well as cases where they were not guilty of a bad shoot. Facts are far easier to convey than psychological aspects of law enforcement.

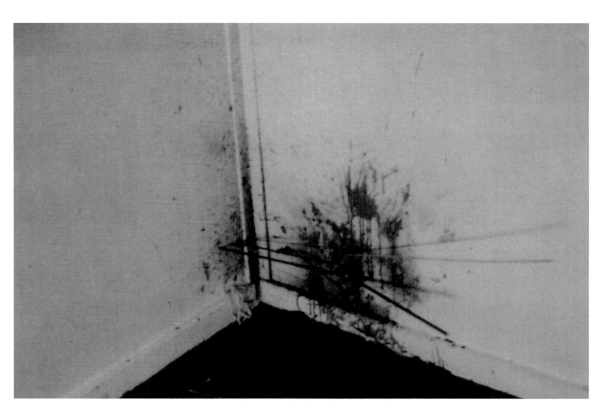

Figure 17-1   Example of the gunshot distribution of impact spatter (GDIS) from a fully automatic weapon. The main male resident's body was found here.

CHAPTER 17

THREE DOWN AND STILL MISSED INTENDED VICTIM

Case 13

BACKGROUND

A garage, attached to a low-rent tract home, had been converted into a one bed-room apartment. The subdivided area included a living room, dinette/kitch-enette, bathroom, and bedroom. At various times three to five people lived in this small space. On the evening of the crime, a drug dealer came to the residence looking for someone with the same last name who had stolen drugs from him. Three firearms—an automatic weapon, a handgun, and a rifle—were used to eliminate four people, although the intended victim was not present. Three adults were killed and one infant injured. One of the men claimed to be an innocent bystander and offered police an account of what happened. The bloodstain pattern evidence was to be used to verify the man's statement.

The principal resident of the home was found in the living room (Figure 17-1), dead from multiple gunshot wounds from the automatic weapon. His wife was found in the bedroom between the foot of their bed and the rear wall of the dwelling. She was cradling her infant at the moment of her death. Revolver bul-lets had entered the mother's body and passed through her to the child, killing the mother but only injuring the child. The woman's brother was found next to the bed, dead from a single rifle shot to his head.

BLOOD SOURCES IDENTIFIED

Photographs were reviewed first, then a visit was made to the crime scene before conclusions were provided in an oral report. Many blood sources were identi-fied from GDIS/entrance (HVIS, back spatter, blow back spatter), and GDIS/exit (forward spatter) were found.

BLOODSTAIN PATTERNS IDENTIFIED

Since the exact position of each victim at the time they were shot was to be used to verify the informant's statements, the whole area around each victim was examined. This included careful review of the ceilings, walls, and floors.

Figure 17-2

Nonfunctioning smoke detector and questionable flyspecks.

On the living room ceiling, separated from the body of the principal male victim, was a scatter of spots with little or no direction of travel but of the correct ABC (appearance, behavior, context) of blood. These were tested with a standard crime scene occult blood test kit, and found not to be blood.

Similar arrangements were found on a nonfunctioning (no batteries present) smoke detector cover (Figure 17-2). Upon removal it was determined that the spots were also inside the cover, i.e., could not have originated from external blood distribution. Further examination showed characteristic arrangements of spots in rows and comma shapes (Figure 17-3). The patterns were identified as flyspeck distributed liquid. The color of the liquid suggested something like tabasco sauce or some other reddish brown liquid. Flies had ingested liquid and carried it to the inside of the smoke detector, where it was regurgitated in an arrangement of spots characteristic of flyspeck blood pattern.

Bloodspatters: The record of a blood drop contacting a surface.

An area in the bedroom showed multiple patterns, including impact spatter (HVIS) and random unclassified **bloodspatters** (spots). An area of convergence was made to show convergence of two impact patterns, one on the ceiling and one on the wall under the window.[1] The bloodstains, seen mostly as streaks, were apparently diluted with CSF and contained bone and hair fragments (Figure 17-4). Attempts to take a reconstruction of the origin further were abandoned for lack of measurable bloodspatters. Looking back from the junction of the two areas of convergence, wall and ceiling, provided an approximation of where the mother was bent over protecting her child (Figure 17-5).

One bloodspatter stain did not fit within the convergence lines of the other stains. This was measured for an approximation of an impact angle to an origin. It was shown to be in alignment with the position of the second adult male body. The stain was photographed, measured, and lifted with fingerprint tape to be sent to the agency's crime lab. Samples of spatters from each of three other patterns also were collected to verify which victim was the blood source of each pattern. The crime

Figure 17-3

Inside smoke detector confirms flyspeck distributed liquid; tested negative for blood.

---

[1]Wonder, *Blood Dynamics*, 42.

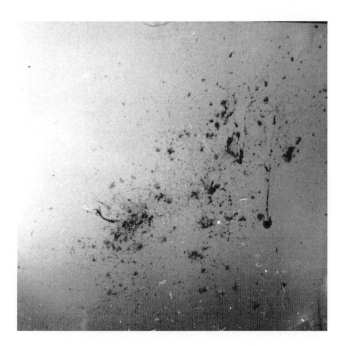

lab later decided not to test the stains because the assailants were known and had left the country by the time of the investigation.

OBJECTIVE APPROACH TO INTERPRETATION

All patterns showed that the victims were killed in the areas where their bodies were found. Each seemed to be in either self-defense or sheltering positions at the moment of the shots.

No specific order could be concluded to the shootings, but the positions of each body suggested that shots were discharged by three different people in quick succession. The positions of the victims and weapons used on each confirmed statements issued by the informant.

ADJUDICATION AND RESOLUTION

The information assisted the investigation in confirming part of the statements made by the alleged bystander.

WHAT CAN BE LEARNED FROM THIS CASE

Not all spots that look like blood are blood. The ABC approach[2] can be useful for improving confidence, but actual testing and verification should be used

Figure 17-4
Often gunshot-distributed impact spatter patterns from massive head injury do not lend themselves to formal reconstruction of the origin. Stains contain brain and bone, and are diluted with CSF.

Figure 17-5
Two areas of convergence in the bedroom can position the mother.

---

[2]Wonder, *Blood Dynamics*, 41–42.

when the evidence is first approached. In this case, the color of the fly specks was the same as the known bloodstains from the victims when viewed. The context was the proximity of multiple spatters from GSW. Behavior could not be verified as no clotting or hemolysis was involved. One should never base conclusive identification of blood just on appearance.

We also can learn from this case that even when perpetrators are known and investigators feel nothing further can be learned from bloodstain pattern analysis, confirmation of eyewitnesses, and positioning of bodies during the assaults can provide useful investigative information.

Figure 18-1   Wall in storage building unit used as a mechanic's shop.

CHAPTER 18

PERFECT STORIES

Case 14

BACKGROUND

A body with distinctive tattoos and scars was found floating in a local river. Detectives were able to identify the victim from arrest records, showing that he had frequently encountered law enforcement. Visits were made to those people known to associate with the victim, but the interviews were conducted as a missing person case. It was soon determined that the last time the victim was seen was at a meeting at a mechanic's garage, located in a storage unit (Figure 18-1). The whole row of storage units was owned by a tow truck driver with one unit used by an ex-convict out on parole working as a mechanic. Three men—the tow truck driver (building owner), his brother, and his hired man—claimed to have been with the victim at the ex-con's shop and to have left the two there. The times given by the three men regarding their whereabouts during the time in question were exactly alike to the odd minute.

The building owner provided the same statements both when the case was presented as a missing person and after it was disclosed that the missing man was actually a homicide victim. On the other hand, when reinterviewed after taking statements from the three men, the mechanic changed his story. This was immediately equated to deliberate lying. The mechanic became the prime suspect, and was subsequently arrested and charged with the murder.

BLOOD SOURCES AVAILABLE

The body was recovered from the river where it had been for some time. No blood sources could be identified in photographs of the body. The shop where the victim was last seen having a confrontation was determined to be the crime scene. The floor of that area apparently had been cleaned with bleach. The exact blood sources could be suggested from the locations of injuries from a beating (to the back of the victim's head) and knifings (to lower back and upper legs) identified at autopsy.

BLOODSTAIN PATTERNS IDENTIFIED

Blunt force impact spatter: Recorded blood drops distributed from a blunt force impact.

Only photographs were used in the analysis. The floor was clean but one wall showed many overlapping spatter arrangements (Figure 18-2). Although the size ranges of the spatters in this case were consistent with those associated with **blunt force impact spatter** (**BFIS**) (MVIS), the patterns in this case did not originate from impact distribution. The arrangements seen on the wall did not fit the criteria for impact (MVIS). There was no convergence. Size distribution was the same throughout the extent of the patterns, and the overall shape was either linear or rectangular.

Some of the stains show a slight up to down direction of travel, and most of them appear to have been directed at right angles to the wall. If these spots were from an impact, there would be clear directions of travel because they were all well above the height of the victim, i.e., the drops had to be going up to reach their position on the wall, becoming more marked the farther from the origin as the whole pattern progressed. The patterns seen were over 6 feet from the floor, and those individual stains further up the wall show less directions of travel and more rounded stains than those at lower positions. The victim was slightly over 5 feet tall. The lack of directions of travel, lack of any apparent convergence, rectangular and linear appearance to the whole arrangement, and the right angle contact appearance suggest that the patterns are all variations of swing cast offs. No impact spatters are identified; Figure 18-3 outlines groups believed to be individual series of cast offs.

The lack of recognizable blunt force impact spatters (MVIS) was not surprising since the assault was primarily to the back of the victim's head. If the victim was facing away from the wall, with his attacker between him and the wall, some impact spatters should have been blocked. No such blockage was identified. If the victim was facing the wall, impact spatters would be directed primarily back toward the wall in a downward motion below the level of the

Figure 18-2

Arrangements of spatters on wall suggest more than one action.

Figure 18-3

Outline of series suggesting separate swings with narrow and broad weapons.

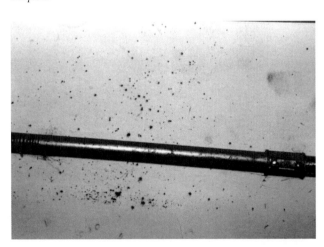

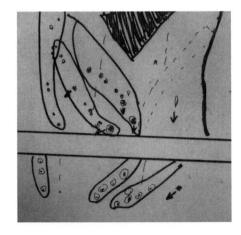

victim (i.e., 5 feet down, or upward above 5 feet). This is illustrated in Figure 18-4. Locating the groups of spatters that represent each swing of a weapon provided an investigative lead. There were both linear and a broader band arrangements. These suggest at least two different weapons, such as a tire iron or crowbar and a two-by-four or ball bat. In addition to the blunt force trauma, the autopsy noted fresh stab wounds. From this we have at least three weapons involved in the assault. The ex-con is not cleared of involvement but it was unlikely that he acted alone if he was an assailant.

A wooden board, which would correspond to the width of one of the cast off patterns, was on the floor with a bloody partial shoe print, but no other stains, (Figure 18-5). This was not available to examine and the crime lab failed to confirm a connection of the print to the suspect (i.e., class characteristics but no individualizing features). The board with the print, however, was found on a bleached clean cement floor with no other blood marks, and no blood source for the shoe print transfer.

OBJECTIVE APPROACH TO INTERPRETATION

The identification of multiple swing cast off patterns suggests that more than one assailant was involved in the beating. Counting the patterns even conservatively, and noting the addition injuries from the knife, strongly suggests that three different weapons were used. Another observation was made once the identification of cast offs was completed. In order for the weapons to distribute cast offs as noted, when all the injuries were to the back of the victim's head suggested he was facing the wall when assaulted. This would project the impact spatters toward the attackers rather than up the wall above the victim's head.

A single assailant may use different weapons, but the presence of three men with identical alibis and the identification of three weapons should have been regarded as worthy of a closer look at the evidence.

The board with a bloody shoe print on a clean floor should alert investigators to a possible staging of the crime scene. If the shoe print were from one of the three men, it is unlikely it would have been left, replaced on the floor after bleach cleaning. Leaving the board with a bloody shoe print suggests an effort to place

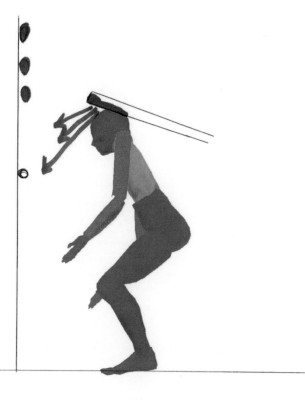

Figure 18-4

Position for assault to distribute cast offs as seen.

Figure 18-5

Shoe print on a board but no blood source remaining.

all the attack on the one person wearing the shoes at the time of the assault.

ADJUDICATION AND RESOLUTION

The public defender that had access to the preceding information decided not to use it. A law enforcement officer for another agency was hired, but identified the stains on the wall as medium velocity impact spatter with direction of travel downward because of arc flight paths. Before the case went to trial, however, the suspect became despondent at the thought of going back to prison and committed suicide in the jail.

WHAT WE CAN LEARN FROM THIS CASE

Here again we have an example where witness statements should not have been relied upon without evidential corroboration. The shoe print was believed to be from the suspect even by his defense attorney. Unfortunately the man had been in prison and was familiar with a code of silence. If the shoe print was his, he was there and could have told what happened. Was he a part of the assault or a bystander? Nothing in the investigation showed a direct motive for the suspect, although there was interview evidence to strongly support a motive for the building owner and his allies. The victim had threatened bodily harm to the man's family from the yard of the man's personal residence.

Another lesson in this case is for caution again to not immediately apply a label of medium velocity impact spatters simply because the arrangement is a bunch of small- and medium-sized spots. To do so here would, and possibly did, result in missing the fact that multiple weapons were used in the assault. Impact spatters merely show where something happened. Cast offs, on the other hand, showed how it happened and that there were more than one assailant involved. Another lesson relative to this case: Training usually involves showing participants that bloodspatters with directions of travel toward the victim, rather than away, are indications that the pattern is *not* impact spatters. Some misunderstanding occurs when the diagram for the trajectory of impacts is illustrated (discussed in Chapter 4). It is erroneous to claim that impact spatters could leave the stains with direction of travel toward the victim due to arc flight paths. The simplified drawing is offered as a representation, modified for space, not the actual flight path of blood distribution. Blood drops do not start out arcing then travel straight. They travel straight depending upon their size and initial velocity, then they slow from air resistance, to fall progressively more by gravity alone, i.e., may arc at the end of the flight path if the path is long enough.

This is why fine and small drops may appear round among medium and large tear drop shapes. In order for an arc flight path to produce spatters with directions of travel toward the victim, blood drops would need to be distributed predominantly up and be large drops. If they were projected up, the victim would have to be very near the wall for the drops to then fall almost directly down.

Another reason to know that the stains on the wall in the garage were cast offs and not BFIS (MVIS) is that those drops forming a rectangular pattern from directed straight up, in order to be falling downward on the wall, would be blocked by the weapon itself.

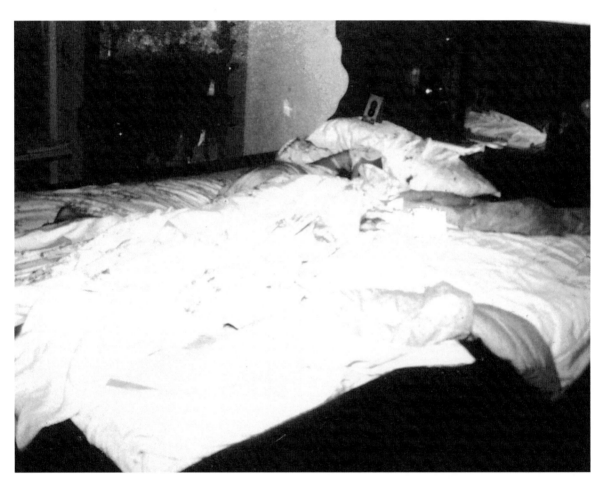

Figure 19-1   Body found in a water bed; mother of an earlier alleged suicide.

CHAPTER 19

FAMILY ELIMINATION

Cases 15 and 16

BACKGROUND

Two deaths occurred a year apart in one family. The first victim was a young man found dead in bed by his father. The manner of death was a single .22 caliber handgun shot to the head, right temple area. The scene was photographed from a distance with limited close-up views, perhaps because the victim was a known HIV positive, terminal AIDS patient. The case originally was accepted as suicide in part based on the father's statements that the young man had been despondent, was alone when the shot occurred, and found soon afterward as judged by the father's interpretation of his body temperature. The father stated he had this knowledge from his past profession as a mortician.

A year later the victim's mother indicated to friends that she suspected her husband of killing their son. When the husband announced he would be out of state for the next few days, the woman arranged to have the locks changed, list their house for sale, and file for divorce while he was gone. The morning the father/husband had flown out of a major airport to the midwest, a cleaning lady entered the house, discovered the woman's body covered with bedding in a heated water bed (Figure 19-1), and immediately called 911. To an investigator not specially trained in the analysis of bloodstain patterns, the arrangement of the body, bloodstains, and three gunshot wounds appeared staged.

Due to the suspiciousness of the mother's death, detectives requested that the DA reopen the investigation into the son's alleged suicide and consider similarities between them as possible homicides.

BLOOD SOURCES IDENTIFIED

No scene visit was possible by the time a bloodstain pattern analysis was conducted. The bedding around the mother's body, her nightgown, and photographs of both cases were reviewed. The analysis, however, proceeded mainly from views in photographs. The cause of death to both victims was GSWs (gunshot wounds) to the head. The male had one shot to the temple area, and the woman had two to the head and one (not mortal) to her right side. All were

.22 caliber. A volume (pool) stain was associated with the son's body. Prior wet stains may have existed but were apparently wiped soon after flowing. A rug that would have been next to the woman's place of collapse next to the bed was found washed and still wet in the laundry along with several towels. The cleaning lady denied washing anything before police arrived.

BLOODSTAIN PATTERNS IDENTIFIED

Since these are to be treated as two homicides but combined in one review, the patterns identified in each will be treated separately, then compared for commonalities.

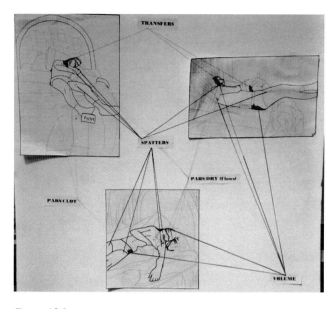

Figure 19-2

Exhibit for court showing flows and transfers contradicting suicide position.

THE FIRST VICTIM (MALE)

One .22 caliber bullet from a revolver was fired into the right temple. Blood flows are seen to predominate and lead to a volume (pool) stain, illustrated with a court exhibit prepared to show flows and transfer (Figure 19-2). There are spatters on the hand, which could suggest a self-inflicted shot. This assumption may be in error if the victim made defensive gestures such as grabbing or attempting to push away the gun at the moment of a shot (Figure 19-3). Such defensive actions could involve recording of residue on the victim's person. Simple presence or absence of spatters should not be sufficient for conclusions of self-inflicted, homicide, or accidental shooting.

Flows seen on the face curved in a manner suggesting that the head was upright at and immediately after the shot, and fell forward before the position suggested by final arrangement of the body. Flows, however, can be misinterpreted due to the depth of field and angle of photography.

Random spatters were seen on the victim's thigh. Cast offs would not be aligned in the directions of travel seen, and no arterial damage was noted. These were considered as probably gunshot impact distribution. No volume (pool) was aligned with the spatters for blood into blood to occur. The face was far removed for a respiratory source to be considered. No arterial damage was noted at autopsy, confirmed with questions to the forensic pathologist from the detective witnessing the autopsy. Confirmation of no arterial damage, since the temple area was breached, was again made by the DA's investigator before conclusions were stated. By the process of elimination, these random spatters were identified as gunshot distributed impact spatters (impact spatters, HVIS). Other spatter distributing events were ruled out.

More photographs would have helped since the angles of those taken were limited. In situations where photographs are limited it is a good idea to provide reconstruction experiments with flow. Since exact blood substance may not be essential in corroborating flow paths in photographs, alternative substances with similar viscosity, such as glycerol or latex-based paint, may sometimes be substituted in these experiments. Latex paint was used in court exhibits to illustrate how viscous fluid would flow from the victim's face.

Figure 19-3

Illustration of hand trying to push away gun at the moment of the shot. This would cause residue on the hand without actually being the shooter.

THE SECOND VICTIM (FEMALE)

THE SECOND VICTIM'S BLOODSTAIN PATTERNS

A series of arcs arranged in flows were seen on the victim's nightgown. They were not consistent with the victim's position in the water bed (Figure 19-4). The arc-like whole pattern suggested the victim was standing upright and then fell face forward immediately after the disabling shot. The gown as seen in the photograph resembled touching blood after the flow started, but it had not been collected soon enough at the scene to avoid contamination. Contamination and investigative transfers prevented conclusions.

One shot was to the right side below and slightly forward of the armpit. This aligns with the hole through the nightgown if the victim was standing bent forward with the right arm extended as if reaching for something. Another shot was to the right temple of the head, and a third shot was to the middle base back of the skull. Because of the arrangement of furniture and walls, the best estimate of position for the first shot would be with the victim standing next to the left side of the bed and bent as if reaching toward drawers under the bed. This position would also suffice for the second shot to the temple. After these two shots the victim would collapse and need to be supported in order to be moved to the bed. The third shot was as the victim was found, i.e., after her body was already positioned in the bed.

Another gun was found in the drawer under the left side of the bed. The possibility that the victim was reaching for a weapon in self-defense was considered. The

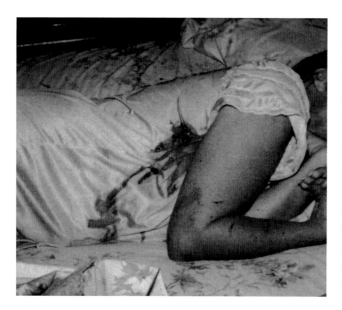

Figure 19-4

Flows on the nightgown suggest an upright position and fall forward with blood flowing for a minute or two before moving to the water bed.

entry to the bedroom was to the right corner of the bed (as viewed from the doorway). If the victim had been in the bed when approached by an alleged intruder, she would have had to immediately sit upright in an unsteady water bed, with the shooter crossing to the left side (as viewed from the shooter's position) to shoot her in her right side without her turning to face the shooter. The movement of the water bed from her sitting up rapidly would make the alignment of this shot very difficult if not impossible. Further, suggesting that the victim did not turn toward the shooter or make any effort to exit the bed was unrealistic, as was the suggestion that a stranger encountering the victim's bedroom would have moved to the opposite side of the bed before firing the gun.

A drip cast off (gravitational stain, passive blood, LVIS) is found on the sheet in a position that would correspond with being under the victim's body. The stain is not smeared and shows the "target rings" or "bull's eye" appearance of partially clotted blood (Figure 19-5); the classification was PABS/clot. Some of the blood substance has congealed but some of the red blood cells are still in the process of sticking together. There is no wound or bleeding injury that would account for the blood drop falling from the victim to the bottom sheet from the position the body was found (Figure 19-6). The drip is also inconsistent with the scenario of the victim sitting up in bed immediately prior to the shots. This bloodstain would be consistent with dripping from the temple wound while the body was being moved to the bed, two or more minutes after the gunshot which exposed a blood source.

Clotting times are not exact, but experiments have shown that limitations can be suggested. Complete clotting with clot retraction and extrusion of clear serum require an hour with normal victims and relatively clean wounds. Partial clotting is not as clear-cut. If movement and jostling occur, well over an hour or as little as a couple of minutes could be involved. The degree of clotting seen in this stain suggests at least a couple of minutes. Body temperature enhances clotting, but with small amounts of blood, like a single drop, major time differences would not be expected.

On a pillow, found on top of the victim's head, was a triangular whole pattern arrangement of spatters identified with directions of travel fanning out from a near point area of convergence. This was GDIS (impact spatter, HVIS) (Figure 19-7). The autopsy identified the wound as postmortem, after death. Since the

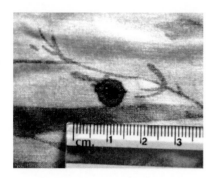

Figure 19-5
Clot in drip cast off.

Figure 19-6
Corresponding drip on sheet but no injury in the area to act as a blood source.

other two shots were premortem this was the last shot fired. The position of the pillows and head locates the shot as fired after the body was arranged in the bed.

The sheets of the water bed were examined. Although the GDIS (HVIS, blow back, back spatter) pattern found on the pillowcase appears consistent with the position of the body, other patterns on the sheets were not in agreement with how the body was found. Other than the drip mentioned earlier, no spatter category stains were identified. The smudges and soaked-in stains were associated with blood transfers and seepage after the body was in place with the sheet arranged over flows, blotting them.

OBJECTIVE APPROACH TO INTERPRETATION

A rule of three may be applied in situations where various interpretations are possible. This rule states that if a conclusion can be arrived at from three unrelated directions or logics it can be stated as a conclusion. Interpretations based on one or two facts are suggested but not necessarily concluded. With bloodstain patterns it is always risky to state as a conclusion anything based upon one pattern. It locks the expert in and jeopardizes testimony if the one interpretation is discredited. The rule of three provides a degree of confidence regarding interpretations. This is especially important when conclusions are based on possible subjective identifications such as HVIS (gunshot distributed impact spatter).

Figure 19-7
Pillow arranged on head when shot to the back of the skull was fired. See Figure 35-2 for the pattern on the underside of the pillow.

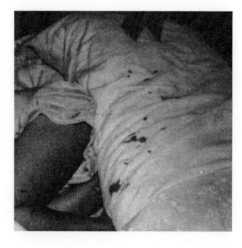

To apply the rule of three we consider all the evidence that is derived from factual statements. Our main concern in this situation was whether or not the two cases suggest a common suspect or are two random occurrences, one being a suicide and the other an interrupted robbery. Five commonalities exist for suggesting a common occurrence with a possible same assailant:

1. Both victims were killed with gunshots to the temple area with .22 revolvers. Such weapons were known to be owned by the father/husband although the barrel/gun combination was not identified. Both victims were known to be uncomfortable around the weapons.

2. Both bodies were "artfully" positioned in the beds after the gunshots, as seen from flows and transfers. The father/husband made a point of his previous experience as a mortician to investigators in both cases.

3. Body temperature was important in both cases and noted specifically by the father/husband as part of his expertise.

4. The male's death did not involve a robbery, however, the scene around the female involved disruption resembling a robbery, although nothing was actually taken. Drawers were up-ended directly over items, which remained as dumped under the drawers. This suggests that the second murder was not robbery-oriented either, thus neither murder actually involved theft.

5. Flows, transfer patterns in blood, and rearranged bedding suggest that both crime scenes were staged.

The bloodstain pattern evidence is important in facts 2 and 5, and are evaluated separately from the other statements. This makes bloodstain patterns additive evidence, not just confirmation for the other information. A review of the bloodstains alone becomes our focus in establishing at least three patterns confirming each conclusion. The first is with regard to artful repositioning of the bodies in each case.

BLOODSTAINS CONFIRMING MALE VICTIM REPOSITIONED IN BED AFTER SHOT TO THE TEMPLE

1. Blood flows across the forehead and down the chin are not consistent with the position in which the body is found, but do coincide with the victim being seated upright and tilted forward at the moment of or shortly after the shot. Having him move on his own to the position arranged on the pillow would be contrary to the pathology conclusions at autopsy. All voluntary action ceased with the gunshot and rigor would not reposition the whole body as found.

2. Spatters found on the male's legs do not show directions of travel consistent with the position of the victim in the bed, i.e., shot not fired while the victim was lying as found.

3. The volume bloodstain (pooling) outlines the arrangement of the right arm in a position not consistent with the fired weapon. The repositioning is suggested.

Bloodstains on the victim's hands are equivocal to either self-inflicted or defensive gestures. These cannot be used as a confirming pattern in the rule of three.

The manner of repositioning with the arm over the head, face tilted, and legs extending to the foot of the bed suggest an artistic repositioning instead of simply pulling the body up onto the pillow.

Separate evidence confirming the repositioning of the male is in the location the shell casing was found. We can thus conclude the male victim was repositioned after he was shot. He could not have done this himself. A second person was present during or immediately after the shot.

The father's own admission was that he found his son soon after the shot because of body temperature, i.e., no other person was present except the father.

BLOODSTAINS CONFIRMING FEMALE VICTIM REPOSITIONED IN BED AFTER SHOT TO THE TEMPLE

1. The flows on the nightgown show the victim upright, bent over slightly as blood flowed, which was inconsistent with the position on her side in which the body was found.

2. The drip cast off (passive stain, LVIS, gravitational) in PABS/clot (partial clot) located under the body shows that the body was placed over the stain at a time period after one of the shots. Since the nightgown would catch blood flows, and the back of the head shot had not occurred yet, the probable blood source for the drip was the temple wound.

3. Alignment of the hole in the nightgown and wound in the side were consistent with the victim standing upright and not as found when the side shot was fired. Since the side wound occurred premortem (autopsy report) and the back head shot postmortem, the temple wound that resulted in death must have occurred second, between the pre- and postmortem shots.

The third shot was to the base of the skull under the pillow, which was placed on top of the victim's head. The position is consistent with the position of the body as found. Since this was found to be postmortem after the victim was arranged in the bed, no logic exists for the shot beyond being part of staging of the scene.

The positioning of the gown as tucked between the victim's legs was pointed out by the District Attorney's investigator, and was consistent with repositioning of the victim in the bed after death. No signs of struggle or defensive gestures were associated with the flows and transfer (compression, contact) patterns. Such actions would be expected if the victim had awakened and encountered a stranger in her bedroom. Again, the position of the arm, legs, and tilt of the head suggest an artistic arrangement rather than just putting the body in the bed and covering with bedding.

Initial investigators attempted to determine height of fall by measuring the diameter of the PABS/clot stain (passive stain, gravitational, partial clot). It must be reiterated here that the fact it was partially clotted is more important than just the fact that it dripped from the victim. In addition, the size of a drip cannot be equated to the distance the drop fell even if the surface material is considered. The composition, ratio of red blood cells in the drop, hematocrit, and overall size of the drop are more important to the diameter of the stain than the distance the drop fell. Unless we know this composition we can't duplicate nor determine from charts of other blood compositions the distance fallen versus diameter. Also, we cannot know the composition of the blood drop unless we know the conditions of the wound at the moment the drop separated. Vessel constriction, jostling motions, and clotting all affect the cohesion of blood, which in turn will determine the size of a drop separating.

ADJUDICATION AND RESOLUTION

The husband/father was convicted of the two murders. The differences between the preliminary testimony of an expert and the revised trial testimony were credited to "viewing the evidence in a new light." In this case the light of knowledge and experience eventually showed through.

WHAT WE CAN LEARN FROM THIS CASE

Looking for specific evidence can lead to subjectivity, i.e., an investigator can find what they want to find when working under time and publicity restraints. On the other hand, verifying that each pattern classification is found or specifically not found is good note taking, but considerably more time consuming. In actual casework there is a fine line between the two.

Reconstruction is not only important in understanding how dynamics happened and illustrating them to a jury, but also useful to check details and provide limits for events. The guideline for reconstruction is to be substantially similar to the crime alleged. This is widely interpreted. The first reconstruction in the female's death involved using a popular 12-inch doll on a solid platform to allegedly duplicate what happened with a startled human on a water bed. Manikins and dolls are definitely useful and favored for jury illustrations, but there is a specific need to consider the actual context in a crime with bloodstain pattern evidence.

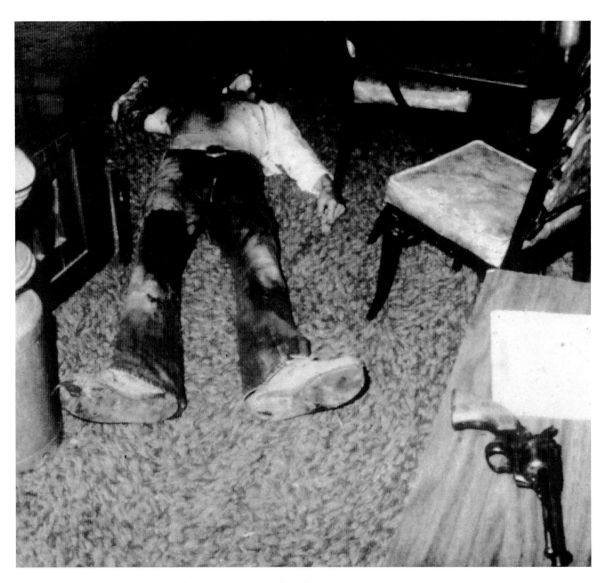

Figure 20-1   Victim's body found inside a mobile home.

CHAPTER 20

A TRAGEDY OF ERRORS IN HOMICIDE

Case 17

BACKGROUND

One evening in a farming area, a man's teenage daughter received a phone call from a married admirer. The young woman did not want this man's attention and told him so. The admirer was on drugs at the time and obsessed with seeing the girl. She complained to her father, who in turn called friends and neighbors to come assist him in discouraging the visit from the man. Later, a 911 call was made by the father claiming that there had been a breaking and entering at his residence. The law enforcement agency with jurisdiction could not respond immediately. Less than an hour after the first call, a second 911 was received alerting dispatch that an intruder had been shot accidentally during a citizen's arrest. Officers were en route by that time and arrived soon after the second call.

The body of the alleged intruder, the would-be suitor of the daughter, was found inside the mobile home residence of the father and daughter (Figure 20-1). Various arrangements of blood and bloodstains were found outside the dwelling, but nothing under or around the body. Detectives in the agency had been trained in bloodstain pattern analysis, but called another expert for confirmation of identifications and interpretation. The true story of events was soon determined and confirmed. The visiting expert merely confirmed what the detectives had seen.

BLOOD SOURCE IDENTIFICATION

In this case the scene, minus the body, which had been removed by the coroner's office, was visited first on the night of occurrence, then followed by review of photographs taken earlier in the evening. The intruder had been shot once in the middle back. No other signs of assault were seen. Tracking of the bullet at autopsy showed it had moved upward and traversed the subclavian artery, located under the clavicle (collar bone), exiting the left side of his neck. This type of injury can be insidious, as shown in the Herman Tarnower death.[1]

---

[1] Twilling, Diana. (1981). *Mrs. Harris*. Harcourt Brace Jovanovich, New York.

The victim quickly bleeds to death without awareness of how serious their situations may be.

Although it is an arterial damage injury, bleeding is into the thoracic (chest) cavity, not as a pressurized projection of blood drops to the external surroundings. In this case the bullet entered the mid-back and traveled almost straight upward. The bullet's exit projected blood onto his head, providing a blood source for his hair, and subsequent blood flows from the neck area and the middle of his back.

Figure 20-2

Vacation taillight showing GDIS and blockage.

BLOODSTAIN PATTERNS IDENTIFIED

A number of patterns were seen, although not all were photographed during initial processing of the scene. When it was noted that some photographs were missing, it was decided that the scene was unrecoverable because rain had fallen in the meantime.

On the side of a separated vacation trailer was a pattern composed of small-, fine-, and mist-sized bloodspatters (HVIS) (Figure 20-2). A blockage transfer pattern (void) of the taillight cover could also be seen. No strong directionality was recognized with the spatters visible around the cover. The blockage, however, suggested the position of the blood source for the spatters at the height of the victim's neck and a location along a path between the mobile home and the separated vacation trailer. An exact origin was not reconstructed from the spatters seen, but a general location of the victim when he was shot was indicated.

Figure 20-3

Hair and skin transfers under daughter's bedroom window.

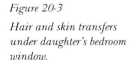

A swipe transfer, probably bloody hair with a skin swipe, identified by the smooth texture and red blood cell density at the lift-off edge, was seen on the mobile home wall across the dirt path from the vacation trailer. This was located under the window of the daughter's bedroom (Figure 20-3).

Drip cast offs (passive staining, LVIS, gravitational) led along a dirt path from under the daughter's bedroom window to the front steps of the mobile home (photograph not available). Three volumes (pools) were noted, one classified as PABS/mix-water, at a standing pool of water at the front corner, near steps to the front of the residence. The other two volumes (pools) were seen on the ground (Figure 20-4). In between them on the ground was a tall circulating lawn sprinkler.

PABS/dry (physiologically altered blood/dry) (flows) were seen on the victim's face inside the mobile home (Figure 20-5), showing directions of travel from the hairline down the face and onto the shirt. PAB/mix-water can be seen on the shirt leading from the neckfold down the front (Figure 20-6), with soak-through to an undershirt.

The breeze in this area was a factor in quickly drying any flows that occurred. Although rain came a day later, at the time of the events the weather was warm, dry, clear, and windy.

The pant legs of the victim's jeans showed straw mixed with blood and blockage patterns for folds in the jeans (Figure 20-7, see page 187). Mud and straw were also seen on the top and edges of his shoes.

OBJECTIVE APPROACH TO INTERPRETATION

Law enforcement detectives are usually very good at developing timelines for crimes. In this case the bloodstain pattern analysis aided in sequencing events and applications of the timeline in questioning the men involved in the death of the intruder. Since the timeline was essential, the patterns defining time/sequence were the focus of interpretation.

The sequencing proceeds with this logic:

At beginning of events
- The manner in which the blood source is exposed
- GDIS, stabbing, bludgeoning with skin breaks, crush, lung puncture
- Events may be outlined by blockages (voids) or defined by absence of bloodstains due to positioning

Middle of events
- Patterns suggesting struggle and/or movement
- Cast offs (drip, swing, cessation)
- Moving transfers (wipe, swipe, smudge)

Later in events
- Patterns suggesting blood changes and alterations
- PABS/dry, PABS/clot, PABS/mix
- Volume accumulation (*pooling*)

PABS/dry (physiologically altered blood/dry): A bloodstain where some degree of drying occurred prior to distribution

Figure 20-4

Volume (pools) of blood, one in water (not shown), and two with straw and tall sprinkler head between them.

Figure 20-5

Flows on the victim's face consistent with victim being upright while blood flowed.

End events

- At, near, or during the time following death
- Volume (pooling) accumulated
- PABS/clot (retracted), PABS/dry

Events extended afterward

- Not involved in crime sequences
- Investigative transfers
- Contamination
- PABS/animal or insect damage

Figure 20-6

Shirt wet, suggesting body was propped against the sprinkler before being moved.

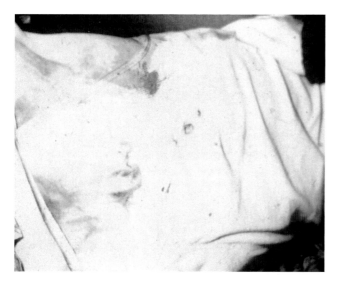

In this case, the beginning of the mist (spray paint affect) pattern at the camp trailer located the beginning of bloodstain pattern events. The blockage (void) by the taillight cover helps position the victim at the moment of the shot. Since the entrance wound would be too low to be associated with the trailer pattern, and the bullet penetrated the shirt and undershirt, thus probably preventing entrance wound spatter (back spatter), we can conclude that the gunshot distributed impact spatter pattern (HVIS) resulted from the exit wound (forward spatter) at the intruder's neck. This is confirmed by the height of the victim and the taillight blockage.

The swing cast offs and drip cast offs show directions of travel back toward the front of the mobile home. We can then conclude that he was shot at the vacation trailer followed by his moving back toward the front of the main dwelling, during which time he bumped against the mobile home. The spacing of drip cast offs along the dirt path is consistent with the victim stumbling under his own power.

The PABS/dry (flows) on the man's face show he was upright for a period of time after blood was flowing. Blood on his shirt indicated that during this time he was exposed to water, and he acquired straw on his pants and shoes. He was placed near the volume (pools) stains for enough time for blood, water, and straw to accumulate. The lawn sprinkler, between the volumes (pools) could provide a prop for the victim, which would include straw acquired on his pant legs and water to the back and neck area of his shirt. The outside wind would dry the flows on his face quickly.

A few swing cast offs and a single drip cast off also line the path to the entry into the dwelling. The frequency of drip cast offs seen on the path is not continued here, showing either that bleeding had almost stopped, and/or the victim was carried in a manner to avoid blood dripping. This would coincide with the victim having died before being taken inside the home.

ADJUDICATION AND RESOLUTION

When confronted with the sequencing, the men admitted that their first 911 call was actually after he had been shot accidentally. He was confronted by the father and neighbors as he looked into the window of the daughter's bedroom. The victim then attempted to run

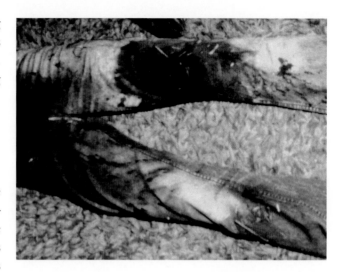

Figure 20-7

Straw on the victim's pants with fold blockages shows that the body was sitting up while bleeding in the presence of water and straw. Placement outside is confirmed.

toward his car, and the neighbor with the gun shot the tires. The man turned back, running along the dirt path toward the field behind the vacation trailer and mobile home. When he observed the large furrows of dirt in the field he abruptly stopped. The neighbor with a gun running behind him ran into him and the firearm discharged. The victim remained outside propped against the water sprinkler while the father called 911. He died prior to the arrival of the law enforcement officer. The second call was made to 911, admitting a gunshot had been fired.

When the men agreed with the detectives that the interpretation based upon the bloodstain evidence was true, the Assistant District Attorney declined prosecution labeling the event accidental death.

WHAT WE CAN LEARN FROM THIS CASE

It has been noted by these and other detectives that solving a case with bloodstain patterns can quickly lead to confessions. The thought goes through the mind of perpetrators that if law enforcement knows so much about the actual events, lying at that point could be detrimental to their case. If the initial investigation deviates markedly from actual events, it gives perpetrators confidence in developing lies. Once statements are signed, it becomes more difficult to obtain admissions of guilt from the individuals.

Bloodstain patterns are free to those who are trained in the use of this form of evidence. Good training and objective application are essential, but well worth the time, cost, and effort to apply.

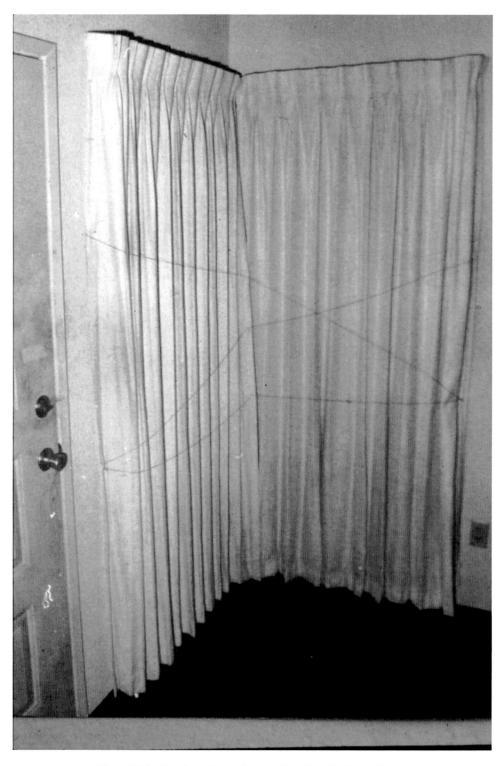

Figure 21-1   The door of a murder victim's residence had been cleaned.

CHAPTER 21

THE ACCOMPLICE WOULDN'T PLEA

Case 18

BACKGROUND

The body of a young adult male was found, just outside city limits, at a refuse dump site that had a reputation for being the venue of various illegal activities. Animals had disturbed the body exterior. After identifying the victim, officers visited his residence in another jurisdiction. It was learned that the victim occasionally had confrontations with his roommate, a teenage boy. Officers arrested the roommate and also detained the roommate's friend as an accomplice. Clothing belonging to the friend was found with bloodstains on a pant leg and surrendered to law enforcement officers by the boy's mother. No bloodstained clothing was found for the victim's roommate.

Detectives quickly developed a scenario in which the roommate and victim had engaged in a final confrontation, and the boy assaulted the victim. When interviewed, the roommate admitted the bad blood between them but claimed the victim had left the house after the fight. Detectives then offered a deal to the roommate's friend, calling him the accomplice. The friend refused to state anything regarding the fight between the victim and his roommate, preferring to confirm the story that a fight had happened and the victim was alive when he left his residence. Early in the investigation attention shifted to the residence. Spots of blood were immediately found on draperies near the front door and it was determined that the door itself recently had been cleaned (Figure 21-1).

BLOOD SOURCE IDENTIFICATION

Material processed for this case included a visit to the crime scene, residence of the victim, access to the complete case file, and open communication with law enforcement and a skilled forensic pathologist. Photographs were reviewed later for confirmation. The autopsy revealed massive head trauma from at least 47 blows with a narrow blunt weapon. Other injuries were noted but none with broken skin to provide a source of blood. Original photographs of the scene were in black and white. Identification was made on the basis of these photographs, but confirmed with a return to the scene and a view of the accomplice's pants. Color prints were taken later. For the benefit of review of cold cases where only black and white photos are all that remain, the initial photos will be shown here. For color views of both the pattern on the wall and pattern on the accomplice corduroy jeans, refer to *Blood Dynamics.*[1]

A bruise was seen in the middle back over the spine (Figure 21-2), but no skin breaks were identified below the scalp. The pathologist pointed out that the blows to the head would immediately raise cerebral spinal fluid (CSF) pressure and that one blow alone had probably caved in the victim's skull to the anterior horn, an area where CSF is manufactured. Figure 21-3[2] shows a frontal x-ray view with a hole in the skull visible at the victim's right rear lobe. The depth of this hole reached the anterior horn of the brain (Figure 21-4). The single blow breaching this area would result in death or coma, leading to death. The victim would not be able to rise and leave the scene under his own power.

Figure 21-2

The bruise in the middle back involved no skin break (i.e., no blood source).

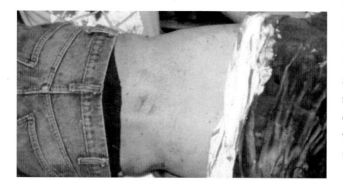

BLOODSTAIN PATTERNS IDENTIFIED

The drapes to the right of the door at the victim's residence showed series of bloodstains arranged as several separate swing cast offs (cast offs). These established that several swings with a bloodied narrow weapon were made to something or someone positioned near the front door of the residence. Photographs did not show the spots well, therefore a print was examined with a magnifier and the dots enhanced with felt tip pen. It was then noted that the 47 blows mentioned by the pathologist were delivered between three positions.

Traditional language states that the number of blows seen in cast offs are equal to the number of blows struck plus at least one. This means that the first blow does not distribute blood drops. In practice several blows may be delivered before blood coats the weapon sufficient for distribution. The patterns on the drapes identify multiple blows with a bloody weapon delivered predominantly horizontal in two positions, but one single blow delivered overhand (Figure 21-5, see page 192).

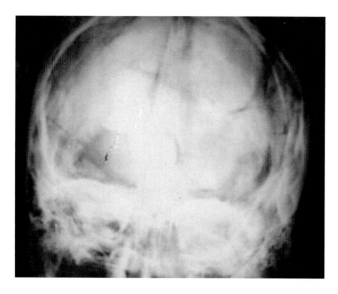

Figure 21-3

Front-on view of skull x-ray shows hole when blow caved in the skull.

---

[1]Wonder, *Blood Dynamics*, 112–113.
[2]*Ibid.*

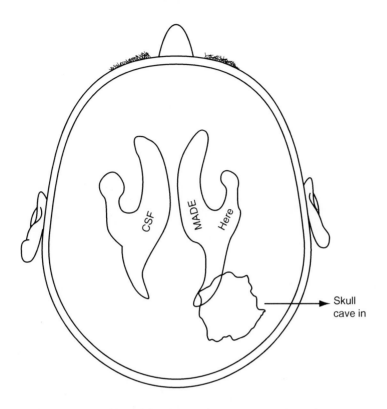

Figure 21-4
Relative area called the
anterior horns of the brain
where CSF is manufac-
tured.

The culmination of the swings drew the eyes of an observer toward an area near the doorjamb. Opening the door showed two stains inside the door frame (Figure 21-6). There was no arterial injury, verified by the autopsy pathologist. Drops distributed during the cast off swings recorded on the drapes would have been in a slanted approach to the door, thus not recorded as streaks inside the frame.

It was concluded from a process of elimination of other spatter events, that the two stains were most likely **blunt force impact spatters** (BFIS) from the head (MVIS). Thus the victim's head was in proximity to the doorjamb during at least part of the assault. The door itself had been cleaned and a rubber mat usually placed at the inside of the door was missing. It was later found with the weapon, covered with the victim's blood and hair, in a trash container nearby.

Blunt force impact spatters: Recorded blood drops distributed when blood drips into a volume of blood.

An interesting pattern was first noticed on the outside of the accomplice's corduroy jeans. A few spatters were seen with the expected color, shade, hue, and saturation of blood. A single round to oval arrangement roughly three inches in diameter was noted near the inside left calf leg position near the knee (Figure 21-7, see page 193). At first the stain was classified as probably a grass or weed stain and ignored in favor of darker red stains, but was reevaluated after a view of the residence.

A three-inch diameter, nearly round, arrangement of pink-colored (noted at the residence but not original photos) fine, mist, and small stains was seen (Figure 21-8, see page 193) on the wall adjacent to the doorjamb where the impact spatters were located. Pink stains may result from various factors at a crime scene:

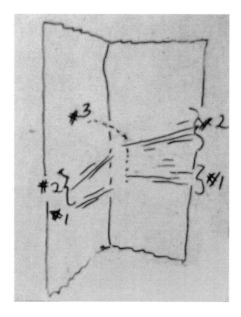

Figure 21-5

Cast offs on the drapes show two positions of the victim during several swings of the narrow weapon.

Figure 21-6

Spatters in the door frame are identified on the basis of elimination.

■ Atomized blood drops scattered over a white background may appear pink when viewed from a distance.

■ Blood diluted with various fluids may appear pink if the red cells are not hemolyzed.[3]

■ Examination with a magnifier showed this pattern to be diluted with a fluid homologous (of the same or greater specific gravity) as blood. Discussion with the forensic pathologist brought out the fluid that fit this pattern, CSF.

OBJECTIVE APPROACH TO INTERPRETATION

The CSF pattern identified on the wall and the accomplice's jeans provides the interpretation that the victim was either dead or in a coma when he was at the residence. Pathology statements verified that he would not be able to leave the scene on his own after his skull was caved in with the blow that broke into the brain. The weapon was later identified as a crow bar. The fatal blow occurred after initial blows which had increased CSF pressure. When the fatal blow caved in the skull, CSF with blood was projected by pressure as mist with small and fine drops. This was recorded in only two positions, on the accomplice's jeans and on the wall behind the victim's head.

The conclusion in the case was that the accomplice, not the roommate, had delivered the blow that caused the victim's death, or, he stood so close to the victim at the moment of the beating that he endangered his own well being. Both positions were shown with artist models in Figures 21-9 and 21-10 (see page 193).

ADJUDICATION AND RESOLUTION

During efforts to sway the accomplice to plead and testify against his friend, no progress was made. When the bloodstain pattern evidence was delivered as part of discovery, the attorneys for the boys agreed to a plea bargain close to that charged.

During the investigation detectives were unable to establish a motive for the accomplice wanting to assault the victim, only for the roommate. A review of the 6-inch thick case file revealed

[3]*Ibid.*, 110–112.

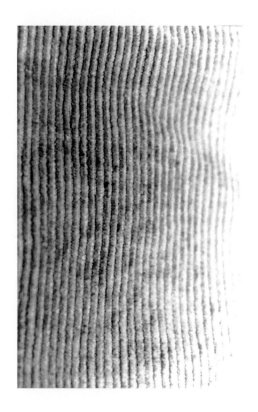

Figure 21-7
PABS/mix-CSF in an impact spatter pattern on the accomplice's corduroy jeans.

Figure 21-8
PABS/mix-CSF in an impact spatter pattern involving mist-sized spatters on the wall to the left of the door (see Figure 32-4 for a color view of CSF and blood distributed under pressure).

a series of events that could explain the animosity between the victim and accomplice.

The victim previously had committed acts to frame others. He was also a drug addict and burglarized homes to feed his habit. A week prior to the victim reported missing, the accomplice became a suspect in a residential burglary. The boy had numerous charges of petty crimes but had never claimed an alibi. During the residential burglary he had an iron clad alibi—he was being interviewed at the police department regarding an alleged assault against his girlfriend. The reason he became a suspect in the burglary was because he had lost his neck chain at the scene of the burglary. When asked where he last remembered seeing the neck chain he answered at his friend's residence, the home of the victim.

WHAT WE CAN LEARN FROM THIS CASE

One lesson may be that when accomplices are adamant regarding not taking a plea against others, there may be a reason. At least this indicates investigators should verify the degree of involvement by accomplices before offering them deals. In this case, the roommate could very well have started the assault but the blow that changed the assault to homicide was delivered by the accomplice.

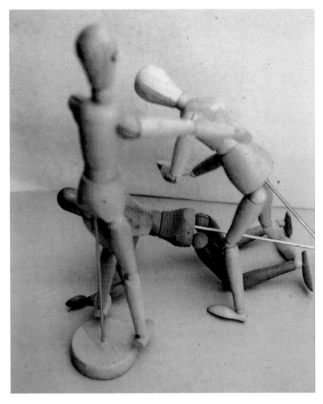

Figure 21-9

Artist model scale of two people position for PABS/ mix stains to be recorded.

An emphasis should also be made on the fact that although beating blows were delivered, focusing on impact spatters (MVIS) would not have provided the information that other pattern types did. The wall photograph from this case was used in many workshops as a practice for string reconstruction. It wasn't until years later, with considerably more case and research experience, that the author realized that the bloodstains being used to locate the origin of the impact weren't impact spatters. They were actually cessation cast offs from one or more of the blows struck to the victim's head. Making a correct area of convergence, which would have made the full string reconstruction unadvisable, can be seen in *Blood Dynamics*.[4] Impact spatters are included in the photograph but were not measurable to use in a reconstruction of the origin. Later workshops pointed out how cessation cast offs can appear to be impact spatters at the scenes of violence. This is another of many reasons to not focus on finding MVIS at a crime scene.

Figure 21-10

Artist model of a single-person assault for blow that caved in the victim's skull.

[4]*Ibid.* 66.

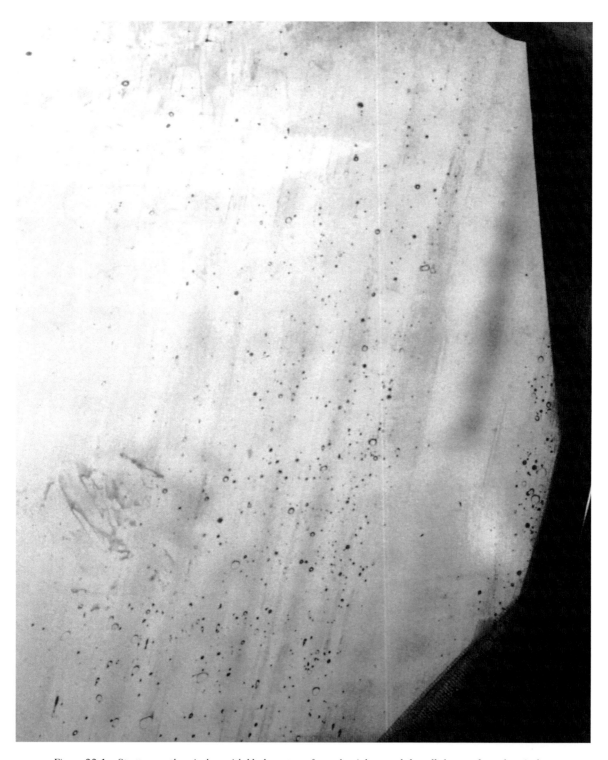

Figure 22-1   Spatters on the window with blockage transfer to the right; seat belt pulled away from the window.

CHAPTER 22

MAGIC BULLET IN ALLEGED DRIVE-BY SHOOTING

Case 19

BACKGROUND

A young man, recognized member of a gang, drove into the entrance of a hospital with his girlfriend in the passenger seat. She had been shot, according to the boyfriend, during a drive-by encounter with a rival gang. Although she was alive when admitted, she died within 24 hours from hypovolemic (loss of blood volume) shock. She provided no information regarding the incident before dying.

BLOOD SOURCES IDENTIFIED

Analysis was predominantly with photographs, but a short time with the crime scene, the vehicle where the victim was shot, was arranged. Medical reports were requested and delivered after viewing the vehicle. The autopsy report was not provided for this case. Emergency room physicians are reliable in noting arterial damage in admission records because this represents the possibility of volume blood loss and imminent fatality. A bullet had entered the right carotid artery and exited the left jugular vein. Since death was not immediate, the heart continued to pump out blood and copious bleeding occurred. Both entrance and exit of the bullet could be recorded in the small enclosed space of the vehicle.

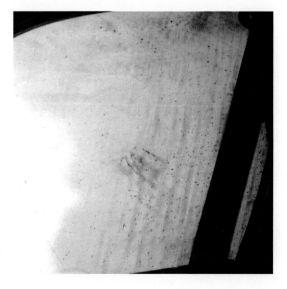

Figure 22-2
Normal seat belt retracted position.

BLOODSTAIN PATTERNS IDENTIFIED

The first pattern identified was a blockage transfer (void) on the right side of the front passenger door window (Figure 22-1). The numerous bloodspatters (bloodstains, spots) outline a strip corresponding with the passenger seatbelt in the retracted position. Pulling the seatbelt forward and locking it into place would shift or prevent the blockage (void) pattern resulting

Figure 22-3

Butcher paper on outside of window to show swipes and wipes on window surface.

Figure 22-4

Spatters seen on the headliner did not photograph well and were small.

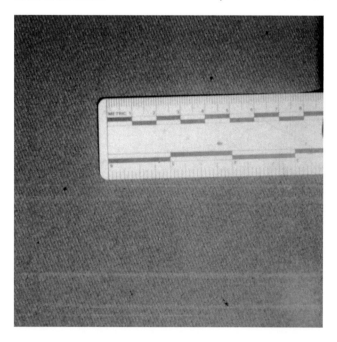

from distribution of blood at the moment of and following the shot (Figure 22-2). Friends and relatives of the victim told police that she was adamant about seatbelt use.

On the same front passenger side window were seen many medium- and small-sized stains lacking directions of travel. Close examination showed these to be the beginnings of moving transfers (swipes and wipes). More than one movement was suggested, i.e. window up to record spatters, then lowered, and raised again. Swipes (bloodspatters on the window smeared) were recognized in an upward pattern as the window was lowered. Wipes were seen with direction of travel downward as blood from the window seal was deposited on the glass while the window was raised. At the evidence storage garage, a sheet of butcher paper was placed on the outside of the door for better viewing of the shifts in direction of the spatters (Figure 22-3).

The pattern recorded on the headliner of the auto was a rough linear whole pattern with similar sized spatters, usually associated with cast off dynamics. Because they were difficult to see in an overall view (Figure 22-4), page hole reinforcement circles were used to locate and highlight each stain (Figure 22-5). Close examination also showed that although the stains extended back into the rear seat area from the front seat, no directions of travel were noted over the length of the whole array. Had these stains resulted from an action such as flicking or cast off of blood drops at the headliner, direction of travel should have been recorded at least on the farthest stains. Any time a spatter pattern involves a discrepancy to classification, alternative dynamics should be ruled out before concluding identification.

In this case there are six possible actions that could distribute blood drops:

1. Respiratory, such as cough, sneeze, wheeze (exhalation, expiration), would require the victim to have damage to the respiratory organs and the mouth or nose aimed toward the back of the vehicle. She did have a blood source exposed in the bullet traveling through her throat. The angle of the head to project stains in the manner recorded, however, was not possible.
2. Swing cast offs would require a greater overhand arc than possible in the vehicle from the front seat. This was considered unlikely, especially in the absence of recognizable directions of travel for the individual stains and the distance into the back seat area covered.

3. Cessation cast offs would also require some directions of travel in this case, more so even than expected with swing cast offs because all blood drops would have to travel from the point near the front seat. Cessation cast offs can and frequently are lacking in directions of travel when they are directed at the surface at or close to 90 degrees. When directed at a surface and moving beyond their origin on the carrier, cessation cast offs should have directions of travel. These, too, are excluded as possible dynamics for the headliner spatters.

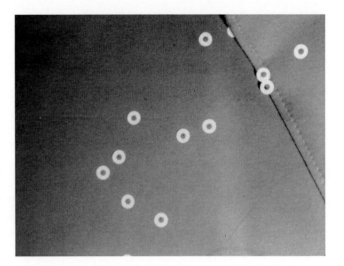

Figure 22-5
Page hole reinforcement seals were added.

4. Arterial damage projected bloodspatters should show some pulsing or undulation to the whole pattern arrangement, but could project stains as seen in a single spurt. This would require the right carotid artery to be aimed at the headliner at the moment of the shot, and some position directing the spurt toward the back seat. If the victim's head were laying against the driver's chest and bent to align horizontally, the gunshot could enter the artery and project spatters at right angles toward the headliner.

5. Blunt force impact spatter (BFIS, MVIS) would require some indication that the victim was struck or hit after blood began to flow. No such event was noted in the medical records, or other patterns in the vehicle.

6. Gunshot distributed impact spatter (HVIS) involved far more than just the concept of a bullet entering a blood source. The force of a bullet at impact also may spread the drop array at GDIS/entrance wound (blow back spatter, back spatter) to create a cone of distributed blood drops. If the cone is intersected by a surface at right angle, the whole pattern may appear linear with a slight arc to the line or lines. In other words a gunshot pattern may resemble a cast off pattern if the shot is fired in a specific position. The victim's head on the driver's chest would be one suggestion of how this could occur.

Other random swipes and wipes were found both inside and outside the vehicle. Some, if not all, of these could be attributed to paramedics removing the victim from the car at the hospital, especially since she was still alive and the artery was breached.

OBJECTIVE APPROACH TO INTERPRETATION

Essential information was immediately available from the bloodstain pattern evidence. The victim was not wearing a seatbelt when she was shot, as shown by the blockage pattern. Since she was adamant about wearing one there is reason to believe the shot occurred before the car was moving. This tended to contradict the boyfriend's statement that they were simply cruising when the rival gang shot at them.

The most significant patterns seen in the vehicle were those on the passenger's window. The window was closed at the moment of the shot, as recorded in spatters, but opened soon afterward and again closed, as recorded in the wipes and swipes. This could indicate the speed at which the boyfriend developed his story. The swipes/wipes combination on the window show that the window was raised before all the blood was dry. By the time the victim arrived at the emergency room the stains would be dry to the extent that smearing would not be expected. Thus the actions of the boyfriend are recorded, not those of the emergency personnel in removing the victim from the car. No sideways smearing was seen, only down and up. No comments were shared regarding whether ER staff noticed the window up or down.

EXPERIMENTS, RECONSTRUCTIONS

Because the pattern on the headliner was of specific interest, two experiments were conducted at the California Highway Patrol Academy near Sacramento. Confiscated vehicles were used with crash manikins to see if the linear pattern on the headliner could be reconstructed with a gunshot. Confirmation of the angle the shot was fired was also available in a bullet hole in the driver's side door panel. The alignment was angled up to down, right to left. The bloodstain patterns on the boyfriends T-shirt (Figure 22-6) showed a blockage (victim's body) and GDIS arrangement (pie wedge with directions of travel away from the point) substantially similar to the reconstruction exhibit (Figure 22-7). Consideration

Figure 22-6

Photocopy of the blockage and GDIS section of the boyfriend's T-shirt. (No cameras were available but a copy machine was in the interview offices.)

Figure 22-7

Reconstruction of boyfriend's (driver's) T-shirt after test firing inside automobile.

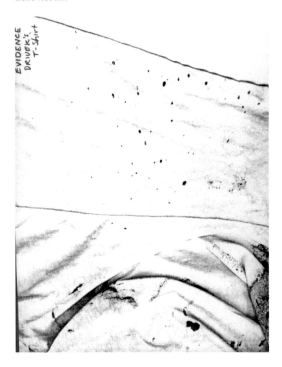

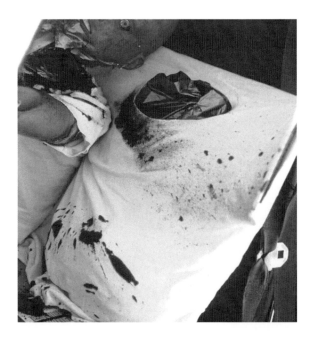

for alignment of the bullet hole in the door, tracking of the shot, and projection to the headliner led to the conclusion that the victim was pulled to the shooter, sitting in the driver's seat, gun aimed up to down against her neck, when the weapon discharged (Figure 22-8). A view from the top of relative positions is in Figure 22-9.

ADJUDICATION AND RESOLUTION

The boyfriend was convicted of shooting his girlfriend. The motive was presented that she was breaking up with him and had chosen to abort his child without his knowledge. She felt an obligation to tell him the truth at what was to be their last meeting.

THINGS TO BE LEARNED FROM THIS CASE

The first thing to be appreciated is the importance of blockage transfer as well as the presence of defined patterns. The most important addition to knowledge of bloodstain patterns, however, is in identifying the gunshot distributed impact spatter pattern on the headliner, which mimicked the appearance of a cast off pattern. This shows the importance of accurate identification of spatter patterns as opposed to labeling from memorized workshop exercise results as pattern match. Identification of evidence must be made within the context of the crime scene where found. Labeling from memorized spatter patterns may be very detrimental to an investigation.

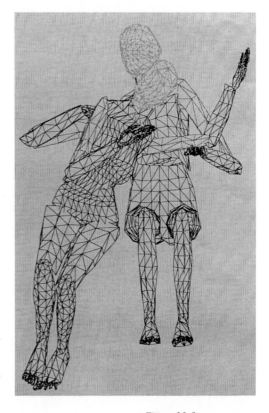

Figure 22-8

Reconstruction of positions using Manikin®. Courtesy of NexGen Ergonomics Inc., www.nexgenergo. com, www.humancad. com.

Figure 22-9

Top view of relative positions using Manikin®. Courtesy of NexGen Ergonomics Inc., www. nexgenergo.com, www. humancad.com.

Figure 23-1 Victim left on dirt berm of side road off a major freeway.

CHAPTER 23

PERRY MASON IS A MYTH

Case 20

BACKGROUND

Three people, two men and a woman, were involved in a drug deal gone bad. They drove to an isolated, rural, unlit cul-de-sac to resolve distribution problems. An argument escalated until one man was beaten, stabbed, and left dead off the road on a weed covered dirt bank (Figure 23-1). Witnesses later the same night informed law enforcement investigators that a victim might be found in the isolated area. The witnesses indicated who was last seen with the victim and pointed officers toward a female and a male suspect. The female immediately agreed to a plea bargain in which she would testify that the surviving male committed the entire murder alone, in return for dropped drug and outstanding burglary charges against her.

Only photographs were initially available for review in this case, along with the interview records from the star witness. The autopsy report was sent later. The main question asked was whether or not the witness was being truthful. The suspect offered no statement, claiming he was high on drugs at the time and couldn't remember anything of the events.

BLOOD SOURCES

The victim had injuries to his head but the main source of blood was from a stab wound to the abdomen with knife tracking down to up, slightly left to right. Defense wounds to his fingers and arms were also identified in photographs and the autopsy report.

BLOODSTAIN PATTERNS IDENTIFIED

Drip cast offs (LVIS, passive stains, gravitational, drip trail) were found leading from the parking area of the cul-de-sac toward the location of the body. The arrangement of drops was described in law enforcement reports as straight but is interrupted by a group of scattered drips, then continued as a widely spaced linear trail: Figure 23-2 shows the number cones of stains. Photography of the scene was not optimum for blood pattern analysis.

Figure 23-2

Numbers indicating location of drip cast offs.

Random spatters were found inside the car, in which the three people arrived. One spatter (bloodstain) is seen inside the door panel where the padding was missing. No photographs of the actual stains inside the vehicle were provided, but notes were recorded with an officer photographed pointing to each stain. Other spatters (spots, bloodstains) were pointed out on the console between the seats and near the ignition switch. The law enforcement officers processing the scene specifically noted that there was no blood on the passenger side: door, seat, glove compartment, headliner, floor, nor exterior on that side of the car.

OBJECTIVE APPROACH TO INTERPRETATION

The drip cast offs (LVIS, passive stains, gravitational, drip trail) show the sequence of the assault.

1. Injury occurred near but not inside the vehicle to expose a blood source with the victim beginning to bleed rapidly.
2. The victim ran toward the dirt bank while dripping blood. Spacing of spots suggests running, not the shape of individual drips.
3. The victim abruptly stopped and struggled, distributing drops at right angle to the road in a scattered arrangement. Since the linear arrangement of drips show he was running in a fairly straight line, we can suggest he was running from an attacker. Since he stopped abruptly, one of two things must have happened. His attacker outran him and got in front of him or there were two people involved when he ran and the second person was in front stopping the victim so that the attacker caught up with him. We know there were two people present, and therefore we can consider their relative positions.
4. Since the victim stopped, as seen by the scattered drip cast offs (LVIS, passive stains, drip trail), he would become an easier target for another attack. When the drips again became linear to the dirt embankment, the spacing between drops increased (became farther apart) but also became more evenly distributed, which suggested that he was then

immobile and being carried. The diminished dripping can be achieved with two people carrying a body via shoulders and legs (i.e., the abdominal injury bleeding inside the body cavity rather than dripping externally).

5. Since the victim was both beaten and stabbed, it suggests that both the man and the woman were involved in assaults. The beating was not fatal but the stabbing was.

After review of the BPE, the statement offered by the female suspect was read. She admitted to being there, and was later seen by witnesses as the person who drove away from the scene when she and the male returned to the city. She explained all the bloodspatters (bloodstains) around the driver's seat as jerking actions by the man. He allegedly had reached across her and shut the driver's side door for her, turned the key in the ignition, and even started shifting into drive. The female further stated that she was 10 feet behind the two men when the accused delivered the fatal stab to the victim's abdomen. The woman denied any involvement at all but was enthusiastic in helping to convict the male suspect.

ADJUDICATION AND RESOLUTION

Perry Mason is a myth. Very rarely is there a case where resolution occurs at trial. Cases are constructed well before trial based on the evidence and trial strategies of attorneys, motions submitted, and the trier of fact acceptance of them. This case was a rare one where resolution probably occurred during trial. Although the defense attorney was quite willing to hire an expert, it was pointed out that results could be achieved with the state's own witnesses. The foundation was set with the autopsy pathologist in which they were requested to pantomime for the trier of fact how a knife would need to be used in order to create the injuries seen at autopsy. Although not totally willing, the pathologist gave a clear performance of the way the knife was held and thrust into the victim.

The law enforcement officers were used to verify that the area had been near black out in the dark, with only lights from the car headlights to see by (Figure 23-3). Facing the headlights would nearly blind a person, while facing away would create large shadows obscuring vision.

The star witness (female) present during the assault was deposed on the witness stand. First she was asked if the accused had held the knife in one hand and the wrench for beating in the other or if one was laid on the ground while the other was used. She was slightly confused but still adamant that she had seen the attack and that the man had done both the beating and the stabbing.

The defense attorney asked the woman to pantomime how she saw the accused actually stab the victim. Her actions were exactly the same as those used by the state's pathologist earlier. She was then asked how she could see this from 10 feet

Figure 23-3
Example of darkness of the area. No overhead lights.

behind the accused in the dark. She could not answer. The jury acquitted the male of murder but did choose to convict him as an accomplice. No charges were ever brought against the female, even though her plea included a nullification clause if she lied on the stand.

THINGS TO BE LEARNED FROM THIS CASE

Although the jury firmly believed the woman did the stabbing leading to the victim's death, the prosecutor declined to charge her. This brings out a common misconception that if the person is convicted they must be guilty and if they are acquitted they didn't do it. Our justice system is not that simple. Had the bloodstain pattern evidence not been presented the client would most likely have been convicted of a crime he specifically didn't do. Resolution at some later time often may not happen, and if it does it may be too little, too late of a resolution. Making use of as much of the evidence as you can initially is far better than reliance on justice at an appeal later.

Too often attorneys, especially defense attorneys, wait until trial is in progress before they consult an expert in bloodstain patterns. This may prevent information reaching the attorneys soon enough to save them time and funds before trial strategy is set.

Figure 24-1   Wall and ceiling bloodstain patterns show the victim's face had to be visible to the assailant.

CHAPTER 24

HIDDEN FACE, BLUNT FORCE ASSAULT

Case 21

BACKGROUND

A teenage boy was accused of beating a woman he had never met. It was alleged that the victim was asleep in a mobile home with a pillow over her head to keep out daylight. The accused came into the bedroom and, without knowing who was sleeping in the bed, beat the victim with a golf club from one of the many sets present around the residence. He admitted he needed funds to feed a drug habit, and had come to the mobile home looking for a woman who owed him money. The beating, however, was to someone unknown to the boy, and no attempt to collect a debt was suggested. The victim survived but had minimal memory of the events.

BLOOD SOURCES IDENTIFIED

All identification was made from photographs of the scene with medical reports to confirm the nature of the victim's injuries and health care after the assault. Injury to the head provided blood for all patterns identified. The victim was admitted to an emergency room of a small community hospital. Bone was removed from the entire side of her skull to allow swelling of the brain. Strikes to the head may not be immediately fatal, but swelling of the brain against the rigid bone of the skull can shut off the blood supply, which results in death at some later time. For this reason any injury to the head should be followed by medical examination, despite feeling fine afterward. In this case the life of the victim was saved.

After initial treatment the victim was moved to another larger hospital for follow-up neurological care. The second facility provided the department chief (medical doctor) to testify regarding her injuries, not the primary emergency room staff. Allegedly under prompting by the prosecutor, the second doctor stated that injury specifically had occurred from a golf club hitting the victim through a pillow. Nothing in the medical reports supported this exactness of testimony.

BLOODSTAIN PATTERNS IDENTIFIED

On the ceiling over the middle of the bed was an area of convergence (point of convergence) showing where an impact (MVIS) had occurred directly beneath. Two impact spatter (MVIS) patterns were seen on the wall (Figure 24-1), directly under the line from the area of convergence on the ceiling.

More important to the question of whether the face was seen during the assault was the presence of a blockage transfer (void), positioned between the two areas of convergence on the wall. This corresponded to the size and location of the victim's head when she was seated near the middle of the bed. To each side of the blockage pattern are moving transfers (swipes) in addition to pie-wedge-shaped spatter patterns (i.e., impact) (Figure 24-2).

Figure 24-2

Closer view of pattern on the wall provides views of patterns from impact, blockage, *and* swipes.

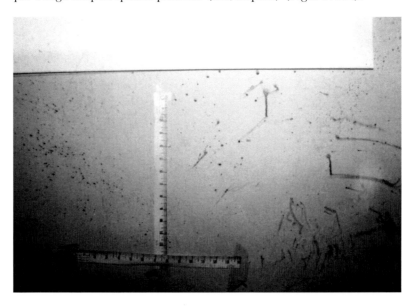

The case of the pillow allegedly lying on top of the victim's head during the assault showed **beading transfers** (contact, compression, simple direct transfers) characteristic of bloodstained hair coming into contact with a recording surface. Some spatters can be seen on the other pillow cases also (Figure 24-3). Conclusion from this is that the pillow was added to the head after blood began to flow.

Beading transfer: When blood forms beads on hair due to natural oils and deposits as a row of dots.

OBJECTIVE APPROACH TO INTERPRETATION

The three areas of convergence (points of convergence) place the beating to the head upright, seated in the middle of the bed, not initially laying on the pillow. The blockage pattern confirms the position of the head during a blood distributing assault. If the blood source was limited to the head, the impacts and swipes would have to be from the head moving side to side as it was being hit. There is no way that the woman's assailant could not have seen her face.

The assault could not have been one of mistaken identity. There was also a question of whether the assault weapon was a golf club.

The hair transfers showed that the victim's head was laid on the pillow after the hair was bloodied, not during an assault. Blood wouldn't bead up along the hair shaft if the head was covered with a pillow during the assault.

ADJUDICATION AND RESOLUTION

The BPE was not considered until the middle of the trial. The trier of fact decided that adding an expert witness would interfere with the court calendar and therefore was denied. The neurologist who was consulted after the victim's injuries were altered by removal of bone stated facts at trial that were not verified for him, yet the defense raised no questions regarding how the physician could identify a weapon and conditions when he had never seen the original injury. The identification of the weapon was beyond the training and experience of the physician. The boy was tried, convicted, and sentenced as an adult. Initial appellate efforts were unsuccessful.

THINGS TO BE LEARNED FROM THIS CASE

Attorneys can benefit greatly from the information regarding bloodstain patterns in their cases, regardless of the other evidence collected. To delay requesting this information can result in serious miscarriages of justice, which may go uncorrected in the legal system.

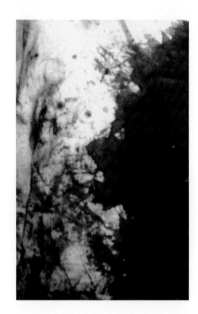

Figure 24-3
Beaded hair transfer pattern shows pillow was added after the assault.

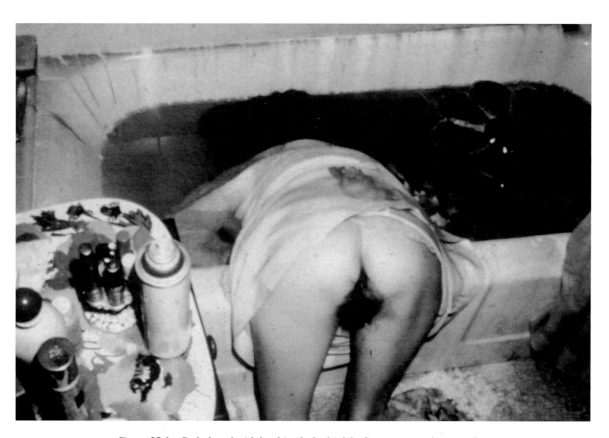

Figure 25-1   Body found with head in the bathtub by firemen responding to a fire.

CHAPTER 25

BODY IN THE BATHTUB

Case 22

BACKGROUND

A fire was reported in a small home in a large metropolitan area. Fire fighters arrived, broke into the home seeking the source of smoke, and found the body of a woman with her head submerged in the bathtub (Figure 25-1). The police were called immediately, and the firemen left after extinguishing the fire. First line officers sealed the scene and called homicide, who in turn called for evidence technicians and the coroner's office. The victim had been vaginally raped in the bedroom, walked to the bathroom, was raped anally and a screwdriver forced into her temple. The residence had been set on fire, but aside from smoke little damage occurred.

BLOOD SOURCES IDENTIFIED

Review was with photographs only, for the purpose of confirming the original findings. No reports or interviews were used in confirmation. The autopsy report was supplied after the preliminary view of the photographs. There was evidence of blood from both rapes and the injury to the temple. The cause of death was stated as heart failure and drowning. The characteristics of drowning such as foam from the mouth and water in the lungs, bronchi, and trachea were absent. Water was in the mouth and the pharynx.

BLOODSTAIN PATTERNS IDENTIFIED

The drip cast offs from the bedroom to the bathroom (no photo available) were as originally interpreted. All photographs supplied were taken in the bathroom. The most significant image was of the bathtub after the body of the victim was removed (Figure 25-2). The evidence is labeled PABS/clot-retracting and PABS/mix-water. Combinations provide a uniqueness that helps individualize a specific bloodstain pattern.

It has been discussed previously that one of the ways to confirm that a substance in photographs is blood is to note unique behavior. A characteristic of blood is the ability to hemolyze. Exposure to water will immediately

Figure 25-2

After initial investigation, photographs, and removal of the body.

cause rupture of the red blood cells so that free hemoglobin (the complex that gives blood the red color) is released into the liquid.[1] Hemoglobin in water is translucent (appears as a clear fluid) like ink or pigmented drinks. Blood in a liquid that does not cause hemolysis is opaque (appears cloudy, muddy).

An experiment was carried out to illustrate the effects of adding water to a tub into which the victim had bled. If blood is allowed to run into a tub of water, such as in Figure 25-3, hemolysis begins immediately. In a few minutes without agitation the whole tub would have turned red such as in Figure 25-4.

If blood clots, however, it changes characteristics and becomes less affected by water. Calcium (chalk) dust was added to anticoagulated blood (ACD anticoagulant) and allowed to set 20 minutes (Figure 25-5). Gentle agitation created hemolysis enough for the red pigment to diffuse throughout the tub although some of the clot material remained.

When the clot forms and firms to the point of retracting, the serum part is water soluble but the blood clot is much less soluble. Calcium was added to previously anticoagulated blood and allowed to sit for one hour. Water was run into the tub and it also was allowed to sit one hour (Figure 25-6). The wound to the temple could not flow into a protected area under the head in the bathtub. Flow would be directly into water if the tub were filled previous to the stabbing with a screwdriver. Clearly the woman was dead or dying, and the pooled blood firmly clotted, when water was added to the tub.

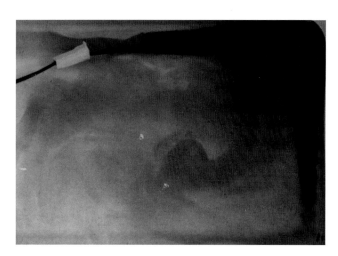

Figure 25-3

Blood slowly flowing into a tub of water. No agitation.

A single spatter (spot, bloodstain) was seen on the back of the victim's leg as found by the bathtub (Figure 25-7). If this spatter was present before she was in the position found, it would be expected to smear while bending the knee to rise herself from the bed and/or lower herself to the bathroom floor. The suggested event for the unsmeared spatter would be during or following the anal rape. Compression of the chest for the rape could aspirate water into the lower respiratory tract. Water was added to the tub after death but before the anal rape.

---

[1]Wonder, *Blood Dynamics,* 114.

OBJECTIVE INTERPRETATION

The conclusion that the blood from the temple wound clotted before the water was added to the bathtub is significant, but the reason was never shared with the author. A suggestion was made that the assailant felt putting the victim's head in water would keep her face identifiable after the fire. Other reasons or no reason may apply.

The finding of drowning was probably not justified from the evidence available to the pathologist. The listing appeared to result from the police statement that the victim's head was found in water. The clot stage before water was added shows that it is unlikely she was alive when water was added to the bathtub. It is reasonable to suggest that the anal rape would have occurred after the head was in water, which would account for water found in the mouth.

Why the assailant added water to the tub after the victim was dead is not known. Speculation could suggest that it was believed that such would protect the head for identification, or perhaps the murder was to be covered by the fire and a conclusion of death from accidental drowning might result.

ADJUDICATION AND RESOLUTION

No further information was provided regarding the final resolution of this case.

WHAT WE CAN LEARN FROM THIS CASE

The main lesson from this as well as all the other cases presented in this section is that there is considerably more information possible from bloodstain pattern evidence than just spatter analysis. Some of the information available has been used in casework, and still more that is science-based physical evidence has never been applied. The potential of the evidence actually applied at this point in time is severely underutilized.

Figure 25-4

Hemolysis from blood exposed to water.

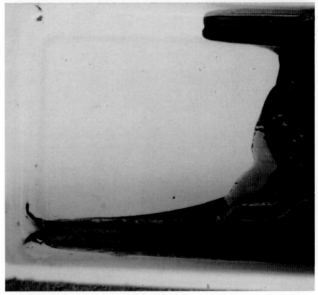

Figure 25-5

Soft articial clot with calcium dust and gentle aggitation.

Figure 25-6

Blood clotted with chalk dust and left one hour before adding water.

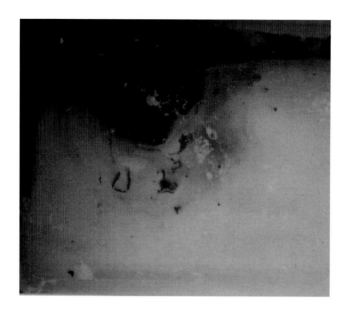

Figure 25-7

Spatter on back of thigh not smeared or smudged. Blood source suggested from anal rape, which could explain aspiration of water into lower respiratory organs.

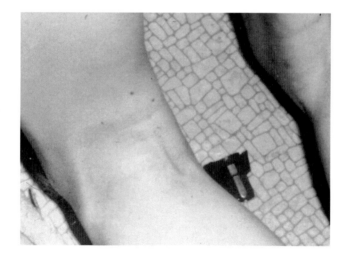

BLOODSTAIN PATTERN EVIDENCE INTERACTIONS WITH OTHER FORENSIC DISCIPLINES

INTRODUCTION

Bloodstain pattern evidence, perhaps more than any other forensic tool, is most effective when used in a teamwork concept. This should be easy to achieve since the traditional training format admits a wide variety of professionals, then divides them among teams to interact. Unfortunately there are occasionally efforts, often by well-meaning budget strapped managers, to limit BPE to a single group of individuals, evidence technicians. Although contributions from these professionals are essential to the whole, "need to know" actually extends throughout the processing of a case. The source of the misconceptions regarding application of BPE is in the belief that this is just blood spatter analysis like fingerprints, sometimes considered a subjective and doubtful science. The only way to deal with this misunderstanding is to show how having knowledge of BPE benefits other disciplines in an objective and scientific manner.

Material for the following chapters was provided by individuals, working in different disciplines, who have sought training in bloodstain pattern analysis in addition to their own professional fields. These are just a few of the many examples encountered between bloodstain pattern analysts and other parts of an investigation team.

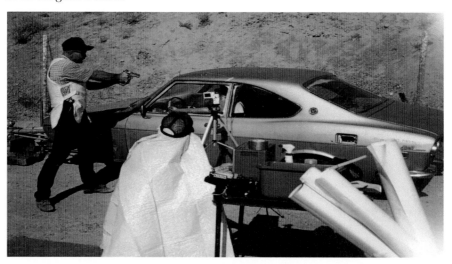

Figure 26-1   Sergeant Reichenberg helps reconstruct a drive-by shooting.

CHAPTER 26

BLOODSTAIN PATTERN EVIDENCE AND LAW ENFORCEMENT

INTRODUCTION

As stated in Chapter 1, law enforcement involvement is undeniably essential in the use of bloodstain pattern evidence. Police are the first, primary, and majority users of information derived from this form of physical evidence. A common theme, however, has been experienced when taking classes with, training, and working alongside the "blue wall." Many law enforcement officers tend to feel that they can only be taught by and learn from fellow officers, sometimes even specific duty officers (i.e., homicide detectives). These individuals may not be the best at working with those peripheral to sworn duty officers.

Training in-house of law enforcement, by law enforcement, for law enforcement alone is within a framework of law enforcement approaches. Science is not a functional concern. A phrase that is repeated often when an approach is questioned is, "but it works." Arguments that using it just to get a conviction are not a measure of it working fall on deaf ears. The reply that is difficult to argue is that if it's accepted in court, that's all that matters. If the approach and explanation can be applied to identify a suspect and that same explanation is accepted in court, many law enforcement officers do not see the advantage of taking the analysis farther nor qualifying it as a true science. Hopefully the case presentations in the previous section will provide an alternative view on how the information can be applied.

Many years ago a graduate from one of the 40-hour classes in Sacramento wrote a thank you regarding his first homicide scene after returning to his agency. The man wrote about his delight when he immediately recognized exactly what happened. He described various patterns and how the suspect had tried to cover up his deeds to no avail, and the satisfaction that followed a complete confession to events exactly as they had been read from the bloodstains. The ending to a long letter, however, was the best thing an instructor could read. The man said that he resolved the investigation quickly, felt good about the resolution, and went home to enjoy his family with a pleasant attitude. This same reaction has been encountered time and again. The evidence can assist law enforcement officers to an extent that far exceeds the gratification of simply getting a conviction in the end. If given a chance it can speed up an investigation, shorten the time and expense in addition to aiding the conviction of the correct suspect, thus also no haunting from cases later.

Benefits are available to all levels of law enforcement including rookies, patrol, traffic, evidence technicians, detectives, as well as those working with incarceration, and internal affairs. Some of the best supporters of the workshop in Sacramento were commanding officers who had started in the ranks and learned the evidence themselves. They appreciated the budget benefits in using evidential leads approaches to bloodstain pattern analysis. Patrol officers who had as little as a one hour lecture tended to do a better job at sealing the scene where bloodstains were found. Rookies did better at not contaminating bloody crime scenes because it gave them guidelines in a trained duty to perform. Evidence technicians are already well aware of the importance but their job becomes more difficult if first line officers have changed the scene. The individuals who have taught the author the most have been detectives. Proper training provides a remarkable economy and efficiency for interviews and case resolution. Following are a couple of examples.

COMMON SITUATIONS, "I DIDN'T KNOW THE GUN WAS LOADED," AND OTHER STORIES

Law enforcement detectives working with evidence technicians processed this case. It was shown to the author during an exchange of information and experiences at Queensland Police Department in Australia. This is such a common situation, however, that law enforcement investigators may benefit from information on how it was resolved quickly. The occurrence was death by firearm (rifle). Investigations in these situations center around the initial determination of whether the death was by accident, suicide, or homicide. The cost and complexity of a homicide investigation is considerably more than for an accidental death or suicide. Being able to resolve what manner of crime happened, and confirm or contradict witness statements quickly, saves time, staffing, and budget.

BACKGROUND

Absence transfer: A pattern resulting from the angle of bloodspatter distribution.

Police responded to a call and found a young woman dead from a gunshot to the head. The boyfriend claimed that he had been cleaning his firearm when the girlfriend leaned forward and playfully pulled the trigger, causing the gun to discharge. His story changed during subsequent interviews but always with something involving the girlfriend leaning across the table and handling the weapon, thus contributing to her own death.

BLOOD SOURCES IDENTIFIED

The only injury was the gunshot wound, but the bullet (projectile, missile) had caused complete destruction of the head.

BLOODSTAIN PATTERNS IDENTIFIED

Several gunshot impact spatter patterns (HVIS) could be seen on the wall (Figure 26-2), ceiling, and table top. The individual spatters contained blood, brain, bone, hair, and CSF. Measuring stains would not have been productive. Efforts to determine areas of convergence were considered, but not completed.

The most important pattern, however, was the blockage transfer. The officers consulted another branch of Australia police and were assisted by Neil Raward, now retired from Queensland Police Department Crime Lab. Officer Raward focused on the blockage pattern instead of concerning himself about the obvious GDIS (HVIS).

OBJECTIVE INTERPRETATION

The height and approximate build of the victim was found from interviewing those who knew the victim well, including the boyfriend. A female officer was located who was of substantially similar height and build to the victim. The volunteer was placed in the seat in front of the blockage pattern (Figure 26-3). The importance of placing a person as a facsimile of the victim was demonstrated later in the effects on the boyfriend.

The original pictures included a view of the ceiling with the ceiling beam reversed. When the original negative was checked it was found that the picture had been printed reversed. The indication from the print that suggested an error in printing was that the pattern on the ceiling included additional blockage and **absence transfers** (void) caused by the beam and angle of

Figure 26-2

"I didn't know the gun was loaded." Australia pre-1988.

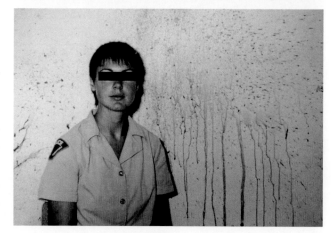
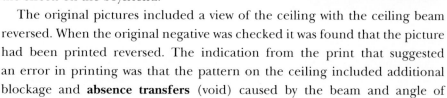

Figure 26-3

Blockage pattern with volunteer helped convince accused to confess that the victim was seated as such when shot.

projection from the gunshot (Figure 26-4). These views of the beams demonstrate differences between blockage transfers (beam obstruction) and absence patterns (lack of spatters due to the angle of projection from the shot, true voids). Using the wall and the ceiling blockage patterns also could confirm location of the origin of the gunshot.

Figure 26-4

Ceiling beam blockage pattern.

Figure 26-5

"Blood everywhere. Accidental death."

RESOLUTION

When shown the picture with the volunteer seated as his girlfriend, the man emotionally confessed that she had been sitting as suggested by the volunteer and that he, not her, had accidentally pulled the trigger and shot her.

Another benefit of bloodstain pattern evidence in quick resolution of a case occurred when police responded to a very bloody scene (Figure 26-5). The parents of the victim were concerned because they had not heard from him. When police entered his apartment they found blood projected all around the bed, and on the floor. The body was found in a short hall near the bedroom. A wrong call could have turned the case into a homicide, but alert and trained officers recognized arterial damage type patterns. Examining the scene identified tissue scraped from the victim's temple on a piece of metal protruding from the bed frame (Figure 26-6). The victim was a drug abuser and an alcoholic with apparent liver damage. Because of the liver malfunction, his blood lacked clotting ability, plus the blood vessels did not constrict properly. A small arterial breach, temporal artery, resulted in his bleeding to death. The death was accidental, and the cost of a full homicide investigation was avoided.

Traffic accident investigation (TAI) has numerous situations for bloodstain pattern analysis. In fact the author has provided one- to two-hour programs for TAI in California and England. As with branches of law enforcement other than homicide and death investigations, the evidence is presently under-utilized. An example of application for TAI occurred in the central valley of California.

CASE STUDY: 37 VEHICLE COLLISION RESULTING IN FIVE FATALITIES

by Dean T. Reichenberg, California Highway Patrol

LOCATION

Interstate 5 (northbound) South of Hood Franklin Road in Sacramento County, California.

INTRODUCTION

On December 11, 1997, at 0707 hours, a collision involving 37 vehicles occurred in a rural area of Sacramento County. The collision resulted in the death of five persons and injuries to 28 others. The collision occurred when visibility was suddenly reduced to between 10 and 100 feet because of low lying fog. Involved vehicles entering the scene did so at varying speeds, resulting in a chain reaction collision. Following the collision, 23 of the 37 vehicles became engulfed in fire. Forty-five separate areas of impact were identified.

ISSUE

One of the prime issues surrounding the cause of the collision surrounded the sequence of collisions that occurred at the north end of the scene and were the first to occur. Those collisions caused two of the involved vehicles to become disabled in the roadway blocking through traffic. One of the involved vehicles was a Chevrolet S-10 pickup (Figure 26-7). The other was a truck tractor pulling a

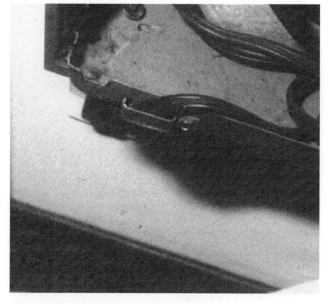

Figure 26-6

The murder weapon. Protruding rough metal on the bed frame nicked the temporal artery.

Figure 26-7

Compact pick up truck involved at beginning of collisions.

45-foot van trailer (Figure 26-8). When these two vehicles became disabled in the roadway they effectively blocked the through path of travel for following traffic. Establishing the sequence of impacts was critical in determining the initial cause of the collision.

Figure 26-8
Truck following pickup.

INVESTIGATION INFORMATION

The driver of the Chevrolet pickup gave a statement following the collision that is summarized as follows: The driver said he was driving the Chevrolet northbound on Interstate 5 in an unknown lane at 60 to 65 miles per hour. He was following a truck tractor pulling a flat bed trailer at a safe distance. The truck suddenly disappeared into a fog bank and he lost sight of the truck. He slowed to 45 to 50 miles per hour and entered the fog bank. He then felt a single impact. He did not know if the impact came from the front or the rear.

Immediately after the collision occurred, the driver of the Chevrolet was found seated in the driver's seat with the driver's side seat belt in place around his body. The driver's seat back was slightly reclined and the driver was bleeding profusely from a large laceration to the right side of his face. The Chevrolet had come to rest in the #1 lane and had sustained major damage to the front and rear. The rear window had been shattered due to intrusion of the cargo bed. The front

Figure 26-9

Damage was minor to the second big rig.

of the Chevrolet was displaced rearward and the rear was displaced forward. The steering wheel ring and column was pushed forward and the dash cowling was shattered.

The truck tractor pulling the 45-foot box trailer was found at rest in the #2 lane approximately 30 feet to the rear of the Chevrolet S-10. It had sustained damage concentrated around the right front bumper and right front fender (Figure 26-9). The right front cowling, right side of the hood, and the right front fender were highlighted by a loosely defined pattern of blood spatter. The spatter adorned the superior surface of the right front fender, up the entire right side of the hood, and on the cowling surrounding the right side of the radiator grill. The blood spatter conisted of individual blood drops in a loose linear pattern recorded on the painted white surface of the vehicle structures. The majority of the stains were generally large, oblong, and aligned along the longitudinal axis with directions of travel front of vehicle to back (Figure 26-10).

CONCLUSIONS

The only identified source of blood that would have been exposed to the front of the truck tractor during the collision sequence was the laceration injury sustained by the driver of the Chevrolet S-10. The drivers of the truck tractors that the Chevrolet S-10 struck were uninjured. The damage to the Chevrolet's steering wheel ring, column, and dash were consistent with the driver's chest and face striking the structures, producing the laceration sustained to the left side of the driver's face. Understanding that bodies in vehicles travel opposite to and in-line with the primary direction of force applied to the vehicle, a substantial force applied to the front and traveling rearward would have to be applied to the front of the Chevrolet to cause the driver to move forward and into the steering wheel structures. That force is consistent with the Chevrolet striking the rear of the truck tractor/flat bed trailer he had followed into the fog bank. Upon striking the trailer, the front of the Chevrolet was displaced rearward while the driver traveled forward and struck the steering wheel, causing the laceration and producing the blood source responsible for the stain found on the front of the following truck tractor.

Figure 26-10
Displacement drip cast offs from the Chery S-10 driver on the big rig.

The impact with the rear of the flat bed trailer disabled the Chevrolet S-10 in the roadway directly in the path of the following truck tractor/van trailer. The following truck tractor struck the rear of the Chevrolet S-10 pushing the Chevrolet forward. Blood dripping from the laceration sustained by the Chevrolet's driver would have been momentarily suspended as the Chevrolet moved forward from impact by the following truck. As the following truck entered the space once occupied by the Chevrolet, the suspended blood drops struck the front of the following truck and were memorialized by the stain patterns observed. The pattern could be called a "displacement" cast off pattern.

From the bloodstain pattern evidence it was determined that the first collision had been the S-10 hitting the trailer in front of it. From that all the other collisions followed. However, no charges were brought against the S-10 driver. It was satisfying to know the full involvement of all three drivers at the beginning of the investigation.

OTHER APPLICATIONS IN LAW ENFORCEMENT THAT BENEFIT FROM BPE

Underutilization of bloodstain pattern evidence occurs with the division of internal affairs. Often officers accepting assignment in this department are based in psychology and sociology. The application of physical evidence might be better received by officers under investigation, their commanders, and the public concerned. Perhaps even more important is the fact that bloodstain patterns can fill in if fellow officers are reluctant to supply details.

The recognition of the existence of bloodstain pattern evidence is now worldwide. An interesting application for law enforcement training is increasingly encountered, using the expertise after retirement. Many law enforcement officers within the author's age group have or are retiring. Some of them are finding work as consultants, for both prosecution and defense. As better trained and more experienced officers enter the field, the qualification level for private consultants will rise, providing a wide choice for both public and private investigations.

An example of active retirement is illustrated by a retired friend in Australia. Sergeant Warren Day is being consulted frequently where bloodstains are crucial evidence. A case example follows.

THE BLOODY WALL CASE

by Sgt. Warren Frank William Day, New South Wales Police Department (retired)

BACKGROUND

Bloodstain interpretation evidence has been used in Australian Courts for many years; however, it was not until one particular case in early 1965 at the Central Criminal Court in Sydney, in the State of New South Wales (NSW), where bloodstain evidence became under notice for its value as physical evidence found at the crime scene. In this particular case a man, Alexander MacLeod-Lindsay, was found guilty for an assault upon his wife with an iron bar and sentenced for 18 years. His appeal later in 1965 was dismissed. In 1969 a Judicial Hearing under Section 475 of the NSW Crimes Act took place, which found that the original jury's verdict of guilt was correct. In 1990, following new evidence on the interpretation of bloodstains from the United States, a second Judicial Hearing under Section 475 of the NSW Crimes Act was granted. As a result of this second inquiry, the first ever granted in Australian legal history, it was determined if the new evidence had been given at the original trial, a jury would have had reasonable doubt, and may not have returned a finding of guilty. Therefore, the judge conducting the second inquiry recommended a pardon, which was granted in August 1990.

In 1985 to 1986, a number of crime scene examiners from the NSW Police received training in bloodstain evidence interpretation. In the early 1990s the Australian Federal Police commenced a Diploma in Forensic Investigations. This is now a national course in which bloodstain evidence training is offered.

In Australia there are limited opportunities for lawyers to obtain advice with respect to the interpretation of bloodstain evidence. Traditionally, it has been a police officer who has had specialist training in crime scene examination duties or someone from a government forensic science laboratory who will provide this evidence. Fortunately, with the increased awareness by the lawyers in recent years, this valuable physical evidence has become accessible. There is now a source of independent advice. For several years the Australian National Institute of Forensic Science (NIFS), based in Melbourne, in the State of Victoria, has maintained a register and can provide information to lawyers looking for a consultant with specialist knowledge.

In NSW usually when a lawyer seeks advice, the case has already been through a lower court. In the situation a police case will be heard by a magistrate, who will determine if there is sufficient evidence to commit a person for trial. If a consultant is engaged, a request can be made to provide a report direct to the lawyer answering specific questions. Alternatively, the consultant may be asked to provide an Expert Certificate as per Section 177 of the NSW Evidence Act 1995, No. 25. The request is usually some months to years after the actual offense has been committed, therefore the consultant must rely on the evidence supplied during the Lower Court hearing such as crime scene photographs, as well as statements from any witness or police officer. As a result of this time delay, the amount and types of evidence supplied can vary. Experience has shown the most important aspect of an investigation is crime scene photography, which can range in quality from good to poor. The latter may possibly make the consultant's job more difficult.

CASE HISTORY

On a warm summer evening in January 2001 an argument developed between two men who rented adjoining rooms in a NSW country hotel. During the argument one male was stabbed in the chest with the heart being penetrated. In addition, there were wounds to the shoulder and face as well as defensive wounds to the hands. During the assault blood was shed in the two adjoining rooms as well as in the hallway. As a result of the stabbing the victim died and his body was found lying on the floor face up in the hall.

The offender was arrested on site and later charged with murder. The usual police investigation commenced, which resulted in the crime scene being examined and photographed. During the course of this examination a number of bloodstains throughout the crime scene were photographed. Statements were obtained from a number of people and subsequently a Police Brief of Evidence was prepared, which included a statement by a police dealing with bloodstains at the crime scene. At the Lower Court hearing, after hearing the police evidence the offender was committed for trial.

Before the trial took place, the lawyer, appearing for the person charged, requested a report dealing with the police evidence relating to the interpretation of bloodstains. Specifically he asked for:

> Opinions on the chronology of events as set out in the Police Brief, concerning blood spatter pattern distribution and where injuries may have been inflicted within the crime scene.

To assist me, a large number of color photographs showing the scene, the offender, and the postmortem as well as a number of police statements and a postmortem report were supplied. The principal police statement was the one dealing with the interpretation of the crime scene bloodstains.

In order to understand the overall scene better I prepared a sketch plan using the photographs and a floor plan from the hotel. On this plan I plotted the position of the bloodstains and body in relation to the two rented rooms. In my opinion a plan such as this is a valuable investigative aid. It not only helps the investigator, but more importantly it helps the jury to understand the crime scene and the location of any physical evidence. Referring to the sketch plan, let's have a look at the crime scene.

As a result of an argument over the loudness of music a fight took place and a person was stabbed, which resulted in his death. When viewing the overall scenes it appeared the occupant of Room 64 was stabbed whilst in his room. One of the stab wounds penetrated his heart. He staggered out into the hallway, moving east, leaving bloodstains on the southern wall. He then crossed to the northern side and moved back in a westerly direction, collapsing at the entrance of Room 54. The person with the knife then went into Room 65 washing the knife in the sink.

Room 64

There were vertical droplets of blood on the bed and floor (Figure 26-11, see positions 10, 11, and 12). On the front of the cabinet appeared blood that may have been arterial (position 13). Arterial blood was found on the exterior (hall) surface of the door (position 14). Unfortunately, due to the "flash on camera"

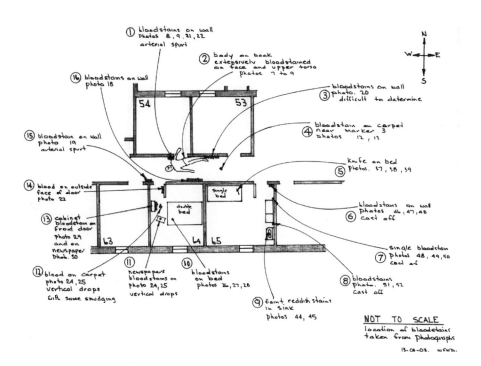

Figure 26-11

Sketch plan of crime scene layout.

technique used to photograph the crime scene, some of the bloodstains on the door had been obliterated and also show up faintly on the cabinet door.

Hallway

This area had five separate bloodstain groups. Two groups—one, on the northeast section of the wall, and the other, on the floor—were undetermined due to crime scene photography (Figure 26-11, see positions 3 and 4). Arterial blood was also found on the north and south walls (positions 1 and 15). However, the most interesting group of bloodstains were the ones on the southern wall adjacent to the body (position 16). I will deal with this group in more detail later.

Room 65

A knife was found on the single bed (Figure 26-11, see position 5). Cast off bloodstains were seen on the east wall of the room (positions 5, 6, and 7). Faint reddish stains, possibly diluted blood, was found in the sink on the east wall (position 9).

Figure 26-12

Showing the bed, cabinet, and door.

Figure 26-13

Showing the hallway looking east.

Figure 26-14

Showing bloodstain on wall between Rooms 63 and 64.

In NSW it is a normal procedure to attach a caption at the base on each photograph, describing its contents. For instance, in captioned Figure 26-16 the bloodstains were identified as swipe, arterial spurt, medium velocity impact spatter, and cast off. Although there is evidence in the police statement relating to measurements, it takes time to sort through this information.

In my view it would be difficult for a jury to understand the location and the importance of the bloodstains in the manner given. I believe much thought should be given to how bloodstains can be clearly identified and then presented in such a manner for a jury to fully understand their contact with the overall investigation.

Figure 26-15

Showing an oblique view of the bloodstains on southern wall.

I have marked up five separate bloodstain groups in the photograph shown in Figure 26-17—three to the left of the measure tape, and two to the right of the tape.

Left Side of Tape

Group 1—Contact transfer

Group 2—Arterial spurting

Group 3—Cast off

Right Side of Tape

Group 4—Cast off

Group 5—Arterial spurting

There may be better methods for indicating the various bloodstains clearly; ringing them is just one way. The main point is to consider how the information is to be presented to the jury in a manner they can understand.

Figure 26-16
Other analysis received by the court.

COMMENTS

Bloodstain evidence can be the most important physical evidence available in the investigation of a crime. Therefore it is important for the whole scene to be recorded carefully to obtain good quality photographs, evenly illuminated, which include a suitable scaling system.

When preparing a Brief of Evidence for court, the bloodstain analyst should give a lot of thought to how the evidence will be presented. It must be done in such a manner that will allow the jury to fully understand its value.

In this case history, it will be noted that some bloodstains were classified as medium velocity impact spatter (MVIS). I wonder if this classification type is correct. I have looked at another case involving a shotgun where high velocity and low velocity bloodstains were identified within the one overall bloodstain group. I believe it would be better to relate the bloodstains to the physical actions, causing them to be formed such as impact, cast off, or arterial gushing.[1]

INCARCERATION APPLICATIONS

Law enforcement officers working in jails and prisons could use bloodstain pattern evidence training in many ways. California Department of Corrections has included the subject in both their basic and advanced investigation

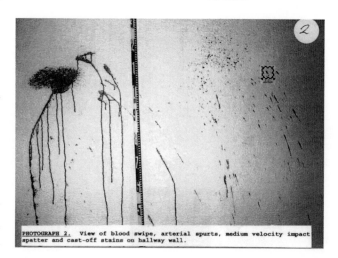

PHOTOGRAPH 2.   View of blood swipe, arterial spurts, medium velocity impact spatter and cast-off stains on hallway wall.

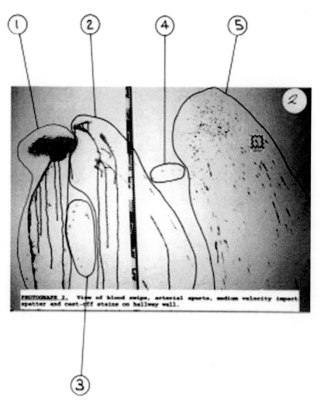

PHOTOGRAPH 2.   View of blood swipe, arterial spurts, medium velocity impact spatter and cast-off stains on hallway wall.

Figure 26-17
Showing circled identification of overlapping patterns.

[1] A note from the author is that as yet, Australia does not have defense discovery. Fairness in their courts would improve if the defense were permitted to view all the evidence collected during an investigation.

Figure 26-18
Arterial damage pattern in an institution.

programs. A simple example of a question that could be answered quickly might be whether an inmate committed suicide or was murdered. The arterial damage pattern seen on the wall in Figure 26-18 is to the left of the commode if someone were facing it and leaning over. Smears, smudges, and transfers confirmed this position for the victim. The question is, was he left-handed or right-handed, and which artery was cut?

The hand that held the knife would have blood on the fingertips but not in the palm. If the victim were left-handed the knife would be held in the left hand. It would then be used to cut the wrist of the right hand. When the artery was breached, projected blood would extend from right to left (i.e., be recorded on the wall to the left). If the victim were right-handed, the evidence would indicate something was not natural about the positions and actions.

PHOTOGRAPHY, WITH COMMENTS FROM EVIDENCE TECHNICIANS ACROSS THE UNITED STATES

Police evidence technicians have their own manuals and guidelines for photography. As a private consultant, the author usually received whatever material is provided in order to locate, identify, and classify bloodstain patterns that were present at the original scene. Some notes may be of benefit within this publication.

As mentioned earlier, if a 35 mm camera is used, color glossy prints are best. According to a sergeant who also teaches photography, either digital or 35 mm are acceptable but a consideration must be made for protection of data with the digital. Micro-chips can be easy to lose, and digital photos run the risk of a charge or modification. SLR cameras have negatives that are easy to store and can provide proof that tampering did not occur. Digital cameras tend to be expensive but they do not run out of film. The entire series of photos for a mass homicide can be taken consecutively and made available immediately. There is no danger for misplaced film or accidental destruction during development.

Another evidence technician, who is a professional photographer on the side, emphasizes that photos be taken at right angles to the evidence both with and without identifying numbers and such. The author adds to this the need for additional views from oblique angles. Head-on flash pictures bleed out blood. Figures 26-19 through 26-20 are all of the same exhibit. The difference is simply the lighting. If photos at different angles are included, at least one of the shots will reveal the bloodstain pattern present.

Another reason for including oblique views is for PABS type bloodstain patterns as seen in Figure 20-22. The three-dimensional characteristic of clot material is not appreciated with head-on right angle views. A 45- to 80-degree angle is better for reflection and details, but a direct 90-degree shot is a must for comparison. Both are essential. Blood and water versus blood and CSF have the same need to include oblique as well as direct photos. Frequently one or the other photographic view is included, but often not both. Comparative photographs are important to reconstruction, measurements, perspective, and interpretation later.

Figure 26-19
Firearms target (.22 round nose, lead bullet).

SUMMATION

A general misconception during testimony for officers is that they must provide for the trier of fact the technical background, or not be considered expert enough to provide investigative information. This is analogous to a traffic officer being required to describe the engineering of an automobile before giving evidence regarding a traffic accident. Hematology, rheology, and non-Newtonian fluid mechanics are not subjects taught to law enforcement officers and should not be required of them. On the other hand using descriptions lacking in science basis and containing out-of-date principles and approaches should not be practiced. The common ground between these two is to define the parts of the whole to be contributed by each department of the investigative team. Bloodstain patterns fit a team work model better than any other form of physical evidence.

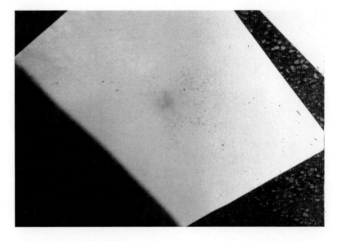

Figure 26-20
Same target as Figure 26-24, but different light position.

Here is where bloodstain patterns fall through the cracks. Law enforcement officers have the experience and exposure to bloody crime scenes. It is within their duty to protect, collect, and interpret the dynamics of crime.

Figure 26-21

Same target as Figure 26-24, but different light position.

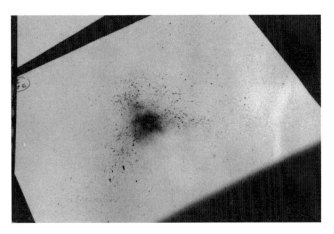

They collect trace evidence, samples for DNA, bottles, beverages, and foods for toxicology, and they describe the context in which these are found. The crime lab has no trouble with this. Bloodstain patterns, however, end up in an undefined area between the law enforcement and the crime lab. Many labs now have nothing to do with the blood spatter analysis [sic], assigning it totally to police work. Law enforcement, on the other hand, is not in the position to define and describe technical science basis for the evidence. What is necessary is for the crime lab to define and support law enforcement with the science principles but to allow law enforcement officers to do what they do best, investigate crime.

The next chapter is an example that the crime lab can be very much involved, if directors appreciate the budget benefits of including the physical evidence as a science subject.

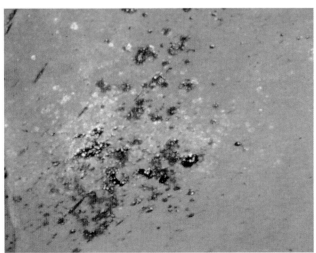

Figure 26-22

Three-dimensional appearance of clot material requires oblique angle and lighting.

Figure 27-1   Experienced criminalists know that interpretation is always important.

CHAPTER 27

THE FORENSIC CRIME LAB AND BLOODSTAIN PATTERN EVIDENCE

**by G. Michele Yezzo, Forensic Scientist,
Office of the Ohio Attorney General,
Bureau of Criminal Identification and Investigation**

TRACE, PROJECTILES, AND SHOE PRINTS

BACKGROUND

During an autumn walk in rural central Ohio a family made the unpleasant discovery of a partially clad body laying at the edge of a creek. The victim, a male, had been shot several times. The deceased was identified and it was determined that he was last seen in a bar in a neighboring city the evening before. Witnesses stated that he had been flashing cash during the evening and later left with a young couple. The male subject was soon located in another city and interviewed. Ultimately a search was conducted of his vehicle and various items of evidence were collected, including a handgun.

BLOOD SOURCE IDENTIFICATION

The deceased had sustained five gunshot wounds to the back and a wound from side to side through his right knee.

BLOODSTAIN PATTERNS IDENTIFIED

No discernable bloodstain patterns were noted at the scene where the body was found. However, examination of a vehicle later obtained from a suspect revealed blood spatters (M or HVIS, Mist, GDIS) in the grille (Figure 27-2) of the passenger side door speaker. Drip cast

Figure 27-2

Spattered stains inside speaker grille in passenger door of suspect's vehicle.

offs (passive staining, LVIS, gravitational stains) (Figure 27-3) were noted on the vehicle floor and carpeting, and blood was soaked into the passenger side front seat. The staining was not apparent upon initial examination of the seat. However, observation and experience suggested that further analysis should be conducted on this area. Surfaces were swabbed and presumptive testing (phenol-phathlein/o-toludine) was positive for the possible presence of blood. Cutting into the fabric revealed visible staining in the sponge underneath (Figure 27-4).

Figure 27-3

Drip cast off staining on carpeting in suspect's vehicle.

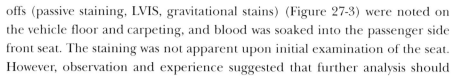

OTHER PHYSICAL EVIDENCE

Scene

During examination of the muddy area adjacent to the body, a partial footwear impression was noted including the letters "bok" (possible Reebok brand). The impression was photographed and cast for later comparison (Figure 27-5).

Deceased

At the time of autopsy, spent bullets were removed from the deceased body for firearms' examination.

Suspect

A weapon and articles of clothing, including a pair of sneakers, were obtained from the male subject. The sneakers appeared to be new or nearly new and were of the Voit brand. The suspect was known to wear Reebok sneakers but none were found during the search. However, analysis of the Voits revealed PABS/mix-water (physiologically altered blood stains), specifically, that human blood mixed with water to form dilute bloodstains on both of the shoes (Figure 27-6).

Figure 27-4

Soaked-in blood staining in passenger front seat of suspect's vehicle.

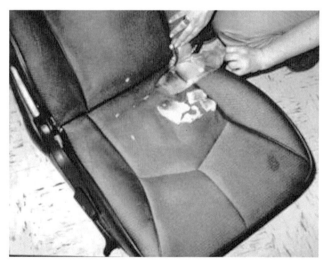

Vehicle

An impression was noted in the grille over the speaker in the front passenger's door (Figure 27-7). This speaker grille was collected as evidence. The impression was compared with impressions made by the lands and grooves of bullets that were test-fired from the suspected weapon and were found to be consistent. Due to the limitation of the size of the impression and the medium that it was registered in, it was not possible to make a positive identification (Figure 27-8).

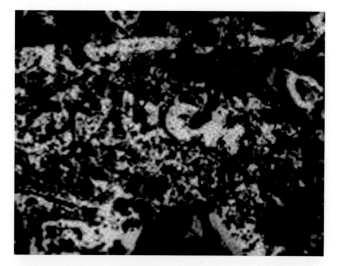

Figure 27-5

Plaster cast of "bok" impression at crime scene.

OBJECTIVE APPROACH TO INTERPRETATION

The identification of spatters (M or HVIS, mist, gunshot distributed impact spatter) deposited in the speaker grille in the passenger door of the vehicle confirmed that at least one shot took place with the victim inside or next to the door of the vehicle with the door open.

The PABS/mix-water in the vehicle (particularly the front passenger seat) suggested that after the incident there was an effort to clean up the vehicle. The PABS/mix-water on the Voit brand sneakers suggested that the shoes were present during the clean up and came into direct contact with PABS/mix-water.

Investigative information was obtained from various sources that this subject "always wore Reebok sneakers." The presence of a partial footwear impression ("bok") in mud at the scene also suggested attempts to destroy evidence of this crime by disposing of his Reebok sneakers.

RESOLUTION AND FINAL DISCUSSION

Confronted with the combined physical evidence, the male suspect confessed. He implicated the female suspect and ultimately they both pled guilty.

Figure 27-6

Dilute (PABS water) bloodstains on Voit sneakers.

WHAT WE CAN LEARN FROM THIS CASE

This is an illustration of the team approach and an excellent reminder of the value of interagency cooperation. Some physical evidence was identified from the scene where the body was found (e.g., "bok" footwear impression).

Figure 27-7

Impression in speaker grille in passenger door of suspect's vehicle.

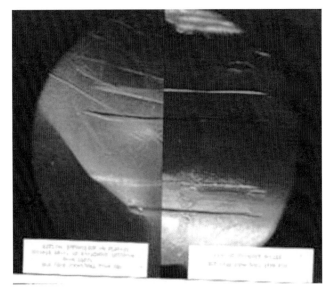

Figure 27-8

Impression in speaker grille/test fire impression.

As a result of the investigation suspects were identified. This led to other evidence, including a vehicle that provided various forms of physical evidence such as the spatters (M or HVIS, mist, gunshot distributed impact spatter) in the speaker grille, drip cast offs (passive staining, LVIS, gravitational stains) on the vehicle floor and carpeting, blood soaked into the passenger side front seat, and a bullet impression on the speaker grille.

We should note that elimination of a piece of evidence by one examination (e.g., the "bok" impression could have been made by a Reebok sneaker, but not by a Voit sneaker) need not necessarily eliminate the suspect. As in this case there was other unexpected evidence in the form of PABS/mix-water stains on Voit sneakers found in the male suspect's possession at the time of his arrest.

It is also a reminder that various forms of physical evidence can be useful and may provide an additive effect. Investigators should not rely solely on any one type of evidence (e.g., bloodstains/DNA) in a case.

LAB CLOTHING EXAMINATION

PATTERNS ON CLOTHING

Any pattern can be registered on articles of clothing being worn during dynamic events. When an individual is in proximity to violence, where blood is shed, garments may acquire stains. This may be true of the victim, subject, and any bystanders.

As a criminalist doing microanalysis (both serology and trace evidence analysis for about 20 years) employed by a fairly large state forensic lab, I normally did not attend crime scenes. That was one of the duties of agents in a separate crime scene unit. As a result, my typical exposure to bloodstain patterns was on items of evidence submitted to the laboratory. This often included clothing. In most cases the required analysis was apparent (e.g., examine the clothing of the subject for possible bloodstains, document the stains, and conduct the serological analyses). However, there were other cases (e.g., a hand-print impression on the shooting victim's shirt) where it was only as a result of an observant investigator that the clothing was even submitted to the laboratory. I credit these investigators, too numerous to be listed by name, with providing me with the opportunity to examine this evidence and correlate it with their investigation. From this I learned things to suggest when training law enforcement officers regarding what we in the lab need from them in collecting and submitting of clothes for examination.

COLLECTION OF CLOTHING

Although this chapter is intended to relate to the examination of clothing articles in the crime laboratory there are some aspects of the collection and handling of these items that bear mentioning.

Clothing from Victim

Often the timely collection and proper handling of clothing articles is extremely important in obtaining results if an examination is to be done later. At the scene of a death investigation careful documentation is necessary, via photographs, before removing the body. Careful packaging and transport avoids obliteration of specific stains in the body bag. Collection should involve consideration of removing the garments at the scene. Cutting through existing holes should be avoided. Powdered gloves should be avoided when handling the body. Documentation of stains caused by handling (e.g., patterns from latex gloves worn) by anyone examining the body may prevent embarrassment later.

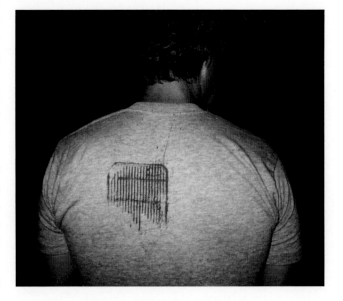

When collecting the garments, at the scene or later at the morgue, package each garment individually in packaging that will allow the garment to breath. Even garments that appear to be dry may mold if packaged in tightly sealed containers or plastic bags (Figure 27-10).

Proper packaging and handling were important in a case of a beating homicide, solved many years after the incident. Among the pieces of evidence in this case were transfer pattern impressions identified on a bed sheet at the crime scene and on the nightgown of the victim. The impressions on the bed sheet included a bloody hand print, partial impressions consistent with the head of

Figure 27-9

Bloodstain pattern evidence training pays back.

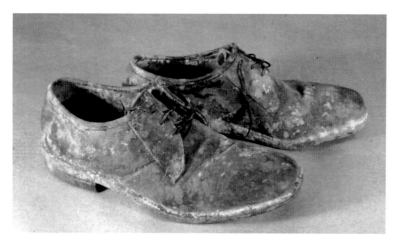

Figure 27-10

Shoes submitted to the laboratory sealed in plastic bag.

Figure 27-11

Enhanced impression on nightgown of victim.

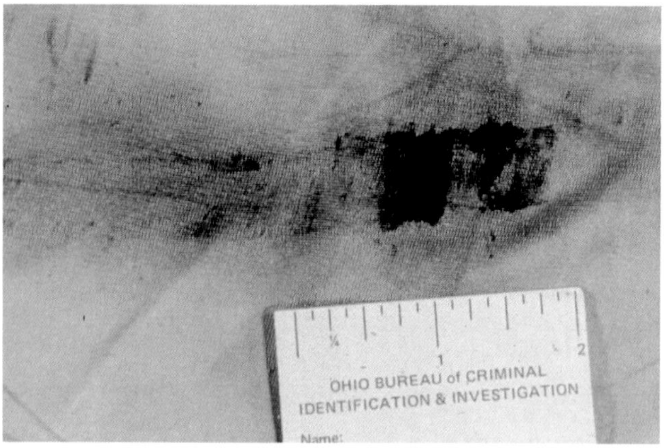

a flex bar, and a linear impression including the letter "N" consistent with the design of the "N" in the brand name CRAFTSMAN. There were also two impressions consistent with the head of a flex bar on the victim's nightgown. One of the impressions, enhanced and documented in Figure 27-11, bore some matching individual characteristics to the head of a tool suspected to be the weapon (Figure 27-12), a flex bar (Figure 27-13), removed from the scene and discovered in the trunk of a vehicle owned by friends of the suspect upon arrival in another state.

Additional problems arise from the perspective of the analysis when the victim survives. Although the primary concern should be medical treatment of the individual, every effort should also be made not to destroy the physical evidence. The potential of investigative transfer increases with each additional individual coming into contact with the person wearing the clothes to be collected.

Clothing from Individuals Other Than the Victim (e.g., "witness" and/or subject of the investigation)

With a living witness at a scene there are additional issues to consider. One should never accept statements made by these individuals unless substantiated by other evidence. Individuals may be mistaken or lying. Ideally the scene should be reviewed for possible bloodstain patterns before interviewing the witnesses. Investigators should be observant of the individual's clothing as worn (e.g., inside out, shorts over trousers, etc.). It is often possible to use the observations to assist in phrasing questions during interviews. If possible, garments should be photographed while being worn.

It is recommended that the outer garments of persons alleged to have been at the scene be collected even in instances where there are no apparent bloodstains.

Other Garments Found at the Primary Scene or Possible Secondary Scene

Garments may be available at the scene (e.g., laying in the general vicinity of the incident, removed by the subject, and placed out of sight or discarded in waste or hampers). These may become important due to statements made later and should be photographed as found before disturbing them.

An elderly man was beaten and stabbed to death in his apartment. During the canvassing of the apartment building for possible witnesses the suspect, whose clothing is documented in Figure 27-14, was found asleep in his girl-friend's apartment. He was wearing the T-shirt over the sweatshirt. However, the jean jacket was found draped over a chair in the apartment. His initial state-ment was that he was asleep and had no involvement in the incident. However, observant officers noticed the apparent blood on the cuffs of his sweatshirt and

Figure 27-12
Head of Craftsman flex bar, suspected murder weapon.

interviewed him further. He soon provided a statement that he was actually present at the scene when another individ-ual committed the act but not actively involved. He explained that he had been sitting next to a table and got blood soaked into his cuffs from the wet blood on the table. He further berated himself for not trying to stop the assault. The subject stated that he did not have his jacket on at the time. However, as noted in Figure 27-15, the cuffs of the jacket bore the same soaked-in staining. The center of the front of the T-shirt and the front of the jacket bore spattered stains (Figure 27-16), and there were cast off stains down the back of the jacket (Figure 27-17).

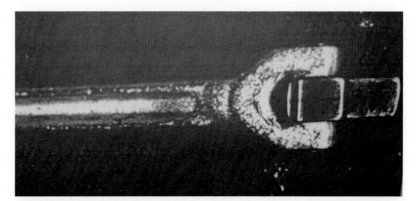

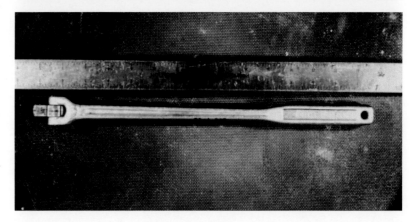

Figure 27-13
Overview of Craftsman flex bar.

EXAMINATION OF CLOTHING

As noted with the preceding case example, clothing, as worn, presents unu-sual contours that should be considered when examining garments for BPE. Interpretation of the stain patterns relates to dynamics and the examiner should consider the motion of the wearer.

Magnification, such as a stereo microscope, can be useful when examining clothing. In some cases it may be necessary to use magnification to distinguish

Figure 27-14

Overview of clothing combination, as worn.

Figure 27-15

Bloody right cuff of blue jean jacket/sweatshirt combination.

small stains on dark surfaces such as indigo blue denim jeans or to determine which surface the staining originated from on thinner fabrics. An illustration of the usefulness of magnification is in Figure 27-18, which depicts a few of the spattered stains on the side of a sneaker. This sneaker was examined during the course of the investigation of a homicide. A teenage boy arrived home from school late, after the bodies of his mother and three siblings had been discovered. Each of the victims had multiple gunshot wounds. The children were covered with blankets prior to being shot. The teen's clothing was requested by the investigating officers, and spattered stains, illustrating at least one impact pattern, was present on his sneakers. The examiner should also be aware and avoid the loss of possible trace evidence that may be on the garments, potentially associated with the bloodstains.

Ideally a view, or photographs of the scene, should be done prior to examining the garments as it may suggest areas of the garment that could have been an available surface for bloodstaining. Such is a case for an incident, where I did attend the scene at the request of the law enforcement agency. The victim, a woman, was in a one-room efficiency apartment and had been bludgeoned over the head while on a mattress that was laying directly on the floor. The impact spatters at that scene were demonstrated on the wall with areas of convergence not far above the level of the mattress, and cast offs were noted on the ceiling. The "witness" claimed that he arrived home and found her as he walked in. He would have had his shoes, socks, blue jeans, T-shirt, and jacket on at the time. However, spattered stains were demonstrated on the lower portion of his blue jeans and on the front and back (cast off patterns) of his T-shirt. There were no stains on his shoes or socks nor his jacket. If his statement was true, there should not have been any pattern consistent with a cast off demonstrated on his clothing. Further if he had been wearing his jacket there would not have been stains registered on the back of his T-shirt. Unlike some common styles these days, the subject's shoes would not have been covered by his pant cuffs.

Pantomiming is often helpful in demonstrating the dynamics of how stains were manifested on garments. Stain patterns on garments are often significant in reconstructing the sequence of events. One example is illustrated by the tire impression on the side of the sneaker of an apparent hit-and-run victim (Figure 27-19). This case also serves as a reminder that it is necessary to have an available blood source before any bloodstains are registered on a surface. It was verified that the tire impression was registered in blood. As a result it would be necessary for the area of the tire to become blood-coated and then rotate back around to the position to register the impression on the victim's sneaker. Since the tire was on the first vehicle that allegedly hit the victim, this presents additional questions as there would be no exposed blood source. In this case, however, there was already a blood source available. The victim had been shot prior to being run over with his own vehicle. That vehicle was later burned. The best blood samples available from the vehicle were taken from inside of the wheel well.

Figure 27-16
Front of blue jean jacket/ T-shirt.

Figure 27-17
Cast off staining demonstrated on back of blue jean jacket.

As mentioned in the previous example, the identification of bloodstains should be correlated with the forensic biologist and DNA analysis. It is especially important in cases where there is more than one possible blood source (i.e., more than one victim, subject claims that it is his/her own blood), or when statements are made that suggest an alternative blood source (wearer allegedly had an altercation with a third party).

It is also useful in some cases (e.g., discarded clothing where the subject claims that it is "someone else's garment") for "wearer DNA" to be done in addition to the analysis of any bloodstains.

Figure 27-18

Spattered stain on side of sneaker.

ADDITIONAL CASE EXAMPLES

Two other examples of garments bearing bloodstains that could be associated with the incident are documented in the following cases. These cases may have involved other bloodstain patterns from the scene that may be mentioned for the sake of clarity but the focus is on the clothing.

CASE 1

BACKGROUND

A woman ran to her neighbor's home claiming that someone had broken in and assaulted her and killed her children. She stated that she blacked out as a result of her injuries.

Figure 27-19

Bloody tire impression on side of sneaker.

BLOOD SOURCES

The woman had numerous small cut wounds in the area of her abdomen that caused volume blood to soak into her shorts and T-shirt. Both of the children had numerous stab wounds including defense wounds. However, when law enforcement officers arrived at the scene it was noted that they were positioned as if they had been put to bed for the night.

BLOODSTAIN PATTERNS IDENTIFIED/ SEROLOGICAL ANALYSIS

Various items of evidence were examined for the case including the woman's clothing and knives removed from the kitchen drawer. A transfer pattern was noted on the T-shirt (Figure 27-20), which was consistent with the design of the set of table steak knives at the scene. Figure 27-21 illustrates the impression on the T-shirt adjacent to one of these knives. A sample from the blade area of one of the knives produced weak chemical reactions indicating the presence of blood. As a result the knife was dismantled and samples from the knife and the knife blade impression were analyzed for blood group and enzyme/polymorphic proteins systems that were currently in use in the laboratory (note: work on this case was done prior to the availability of forensic DNA analysis). The sample from the impression was consistent with the children's, not the mother's blood. The sample from the knife blade was consistent with being a mixture of both.

Figure 27-20
Bloody knife impression on suspect's T-shirt.

ADJUDICATION AND RESOLUTION

The woman was unavailable for any interview as she had been checked into a psychiatric facility. Correlation of the serological analysis with the bloodstain pattern examination assisted in providing information about the sequence of events on this case. The presence of the knife blade impression in the children's blood, its position and orientation, was consistent with the knife being adjacent to the woman's blouse at some time after the blood had been shed by at least one of the children and before the woman sustained her injuries. It was suggested that it may have been against the garment while she was positioning the bodies in bed prior to stabbing herself.

After approximately 15 years this case went to trial and the woman was found guilty of

Figure 27-21

Bloody knife impression on suspect's T-shirt with suspected murder weapon (found in kitchen drawer).

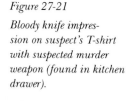
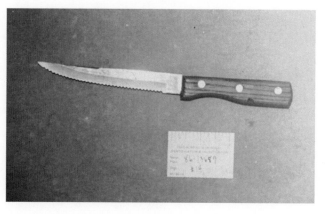

manslaughter. Statements made by her to her clergyman were later made available and reflected that she believed that "… it was her right. They were her children."

WHAT CAN BE LEARNED FROM THIS CASE

This case is an illustration of the value of transfer patterns and correlation of the analysis of serological (now Forensic Biology/DNA) analysis with the BPE. It further demonstrates the value of a good crime scene investigation. Had the knives in the kitchen drawer not been collected and submitted even though they were not apparently bloody a great deal of evidence would have been lost, since the "witness" statement indicated that something else was used.

CASE 2

BACKGROUND

A 911 call was received from a residence that two people had been assaulted. Upon arrival emergency personnel and law enforcement officers discovered a male and female in bed. The male was already dead. He was laying with his head off the edge of the bed and had blunt force trauma injuries to the head and genital area. He also had a lug design impression on his back. The female was still alive but severely injured with blunt force trauma to the head. She was life-flighted and survived in a permanent vegetative state.

The caller was the female victim's estranged husband. He stated that he came to the residence to pick up some things and found the victims in this condition, thus called for an ambulance and remained at the scene.

BLOOD SOURCES

As noted, both of the victims had severe trauma to the head. The male victim also had injury to the genital area.

BLOODSTAIN PATTERNS IDENTIFIED

Impact patterns were noted on surfaces. Due to the position of the male victim's head off the edge of the bed, blood into blood had occurred on the floor beneath. Overhand cast off patterns were noted on the ceiling area. The husband's T-shirt and blue jeans were collected for submission to the laboratory with a request to "type the blood" on the blue jeans. The investigator further stated that no examination would be required of the T-shirt since they saw nothing on it at the time of collection. Due to the amount of available blood at the scene, the investigators concluded the subject either removed his shirt during the assault or had another shirt on that they had not yet located. However, for the sake of thoroughness because of their training they collected the shirt anyway. Although there was minimal staining on the front of these garments, cast offs were noted down the left side of the back of the T-shirt (Figure 27-22) and the blue jeans (Figure 27-23). It is suggested that the subject may have pulled the bedding up to gain better access to his victims and inadvertently, or purposely, shielded himself from the blood during the assault.

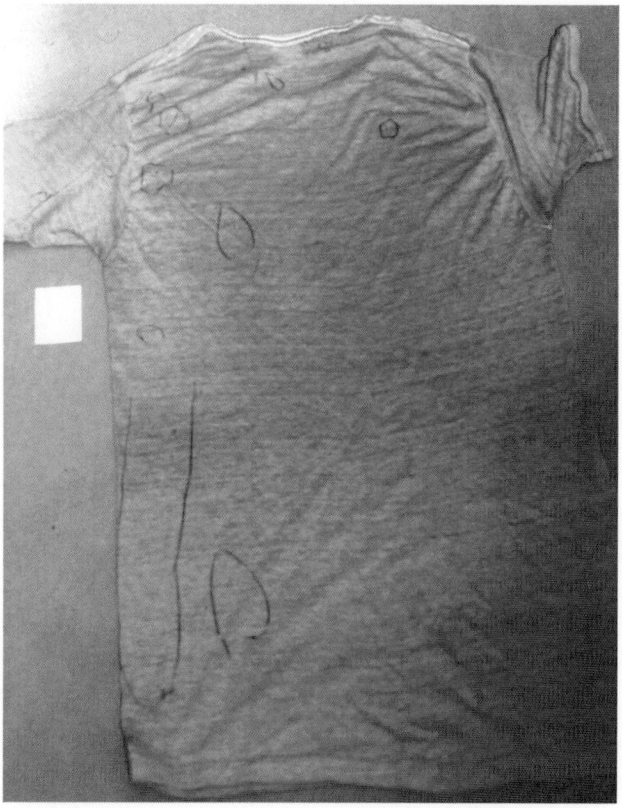

ADJUDICATION AND RESOLUTION

When the results of the laboratory examination were shown to the subject, he provided a statement and pled guilty to murder and attempted murder. The weapon used was a four-pronged lug wrench that was still present at the scene.

WHAT CAN BE LEARNED FROM THIS CASE

Cast off patterns can provide investigative information that is valuable when interviewing witnesses. In this case the value could have been lost due to the misconception that the T-shirt had not been worn during the assault and the delay of submitting the evidence to the laboratory while waiting for the results.

Figure 27-22

Cast off staining demonstrated on back of suspect's T-shirt.

Figure 27-23

Cast off staining demonstrated on back of suspect's blue jeans.

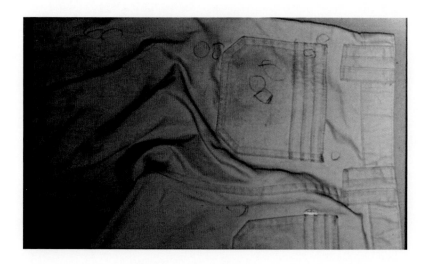

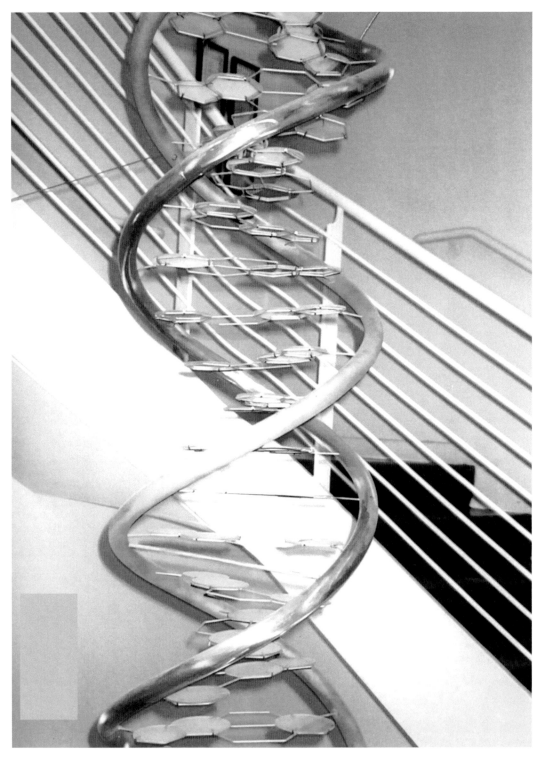

Figure 28-1   DNA molecule in steel and stained glass by Roger Berry in the Life Sciences building of the Davis campus of the University of California.

CHAPTER 28

DNA AND BLOODSTAIN PATTERN EVIDENCE

by Christie T. Davis, Ph.D., Helix Analytical, Inc.

Anita Y. Wonder and G. Michele Yezzo have long impressed upon me their belief in integrating bloodstain pattern analysis with other means of testing physical evidence in order to strengthen the value of each. Cases where bloodstain pattern interpretation and DNA analysis were used in concert provide such a corroborative opportunity. Discussed next are some details from a case where both types of analysis were applied.

BACKGROUND

Two couples were out hiking. They came back to their car to find it had been vandalized by five men present at the scene. Of the two couples, the young men were violently subdued and the young women were raped. Three suspects were developed and all were charged with rape. Included in the samples tested for DNA were apparent bloodstains on items of clothing, and rape kits obtained from both female victims.

Understanding the properties of DNA and the scientific principles of testing and detecting DNA is important in deciding what evidence to collect at a crime scene, knowing what to do with that evidence, and understanding how to interpret the data after testing. This means going back to the basics of biology, biochemistry, and genetics. The aspects of DNA that are required knowledge for interpretation of test results encompass many topics:

- Forensic use of DNA's biological functions
- Physical portions of DNA used in forensics
- Biochemical properties of DNA
- Biological evidence useful for forensic DNA typing
- Replication of DNA by the polymerase chain reaction (PCR)
- Fluorescent chemistry used for DNA detection
- Problems/artifacts that can impact use of DNA in forensics
- Chain of custody of evidence

In this case, the biological fluid in the questioned sample was blood. Blood is a complex mixture of many components including plasma, red blood cells, and white blood cells. The white blood cells are the only source of nuclear DNA in blood. DNA is a double-stranded structure that carries the genetic information required for life. All cells that contain DNA have identical copies. The DNA used for typing results in this case is found in the cell's nucleus. Nuclear DNA is linear and packaged in tightly coiled structures called chromosomes. Each cell contains 46 chromosomes found in sets of two. There are two copies of chromosome 1, two copies of chromosome 2, and so on. One from each set is inherited from the mother and the other one is inherited from the father. Forty-five of the chromosomes are similar between men and women, the forty-sixth chromosome determines gender, X for female and Y for male. By definition, all nuclear DNA together is called the human genome. The genome carries a vast amount of information, only a minuscule amount of which is useful to forensics.

The objective of using DNA analysis in forensics is to differentiate one person from another. Because we are all human and all contain the same essential requirements for life, the vast majority of DNA is identical from one person to the next. For example, the gene coding for insulin is the same for each person. Although this is interesting, it has no value in forensics because it does not help distinguish one person from the next. A small amount of DNA (about 0.1% to 0.3% of the entire genome) has no known purpose and over generations has genetically drifted enough to be discriminatory, meaning individuals can differ genetically from each other. This genetic difference is called polymorphism, defined as variation within a group. Every part of DNA human genome has the same susceptibility to mutation (genetic drift), but the portions required for functional life can be severely affected by mutation. Polymorphic areas have been able to sustain mutations without harm because they do not affect life required functions. Polymorphism is useful in forensics because it allows for inclusion and exclusion of individuals as possible contributors to an evidence sample. The greater the number of these variable regions that are tested in an evidence sample, the greater the percentage of the population can be excluded as possible contributors, which is the primary goal of DNA testing in forensics.

Once samples at a crime scene have been identified and properly collected, DNA typing can take place. By far the test of choice is fluorescent detection of short fragments of DNA termed short tandem repeats (STR). Tandem repeats can be found throughout the human genome. The fragments derive their name from repeating the base units that sit next to each other on the genome. The number of bases in a single repeat unit can range from two to seven bases. The number of repeating units in tandem can range from single digits to hundreds. The FBI study group decided that the forensic test used by U.S. crime labs should contain 13 different tetra (4 bases per repeat unit) repeats.

The "di" repeats (two bases per repeat unit) and "tri" repeats (three bases per repeat unit) created too many artifacts that could cause problems in interpreting the test results. Repeat units larger than four were more stable, but the fragments of DNA started to get too long. The concern was two-fold; longer pieces of DNA degrade (break down) more quickly, resulting in loss of data, and the system chosen for the test worked accurately only within a certain size range. Use of "tetra" repeats kept the artifacts to a minimum and kept the base pair size within the fragment size range required by the test. The 13 different tetra repeats can also be called genetic markers or loci (locations in the human genome). These genetic markers make up the defined U.S. Combined DNA Indexing System (CODIS). CODIS consists of the state and national data banks for convicted felon DNA profiles and profiles of forensic samples from unsolved cases.

How are these genetic profiles generated? First, DNA is obtained from cellular material through a biochemical process. Once DNA is available for testing, only the 13 markers are targeted in the test through a process called polymerase chain reaction (PCR). PCR is sometimes compared to photocopying. The 13 markers are biochemically recognized and millions of copies are made of each fragment. In the process of copying those small pieces of DNA, a fluorescent tag that emits a blue, green, yellow, or red wavelength is attached to one end of each copy to allow detection of the DNA pieces. The length of the fragments and the color of the tag differentiate one STR fragment from another and are the basis for the final typing result. Once the size and color of each fragment are recorded, the answers obtained from a piece of evidence can be compared to the answers obtained from suspects and victims for exculpatory or inclusive purposes.

DNA obtained from evidence samples is considered to be of unknown source and unknown type. This means the person or persons who left the sample is not known and the genetic answer is also unknown until tested. Once the genetic answer is established by DNA typing, the results from the evidence can be compared to genetic types from known sources, called reference samples. Reference means the originating source is known (for example the victim or defendant), but the genetic answer is not known until tested. The results of the evidence sample can be compared directly to those from one or more references for inclusion or exclusionary purposes. If a reference sample has a DNA profile that differs from the genetic marker set obtained from the evidence, that reference is said to be excluded as a possible contributor of the DNA found in the evidence. If the genetic profile obtained from a reference sample has the same genetic marker set obtained from the evidence, that reference is not excluded as a possible contributor of the DNA found on the evidence, and a statistical weight is then placed on the inclusion.

Statistical calculations are based on the genetic frequency of the profile produced from genetic frequencies contained in established DNA databases, and the expectation of finding an unrelated person who has that same profile. The statistical value is based on probability to help determine the rarity of the profile produced from the evidence based on testing unrelated individuals. These numbers are arbitrarily assigned rather than an integral attribute of the evidence sample.

DNA SOURCE

In the case at hand, DNA profiles from two of the suspects were consistent with DNA obtained from semen taken from the female victims. The DNA profile of a third suspect was excluded as a possible contributor to the semen but his DNA profile was consistent with DNA results obtained from apparent bloodstains found on a clothing item from one of the female victims. Of particular interest was an apparent stain on the inside of the bra. The stains are called "apparent" because the type of test used to look for the presence of blood in blood-like stains detects oxidation reactions that can occur in stains other than blood, and thus a positive result means possible blood. Two analyses were done on the sample from the bra. A bloodstain pattern analyst examined the spot to determine the bloodstain pattern and the county crime lab analyzed the spot for a DNA profile at the 13 genetic markers described earlier.

PATTERNS IDENTIFIED

Figure 28-2

Rape victim's bra with a drip cast off smeared on the edges of the under-wire.

Since the apparent blood was found on the inside of a bra, it was important to ascertain whose blood that might be (owner of the bra or someone else), determined by DNA typing, and how it came to be there, determined by bloodstain pattern analysis. Before the crime scene was properly processed (very important in maintaining chain of custody), the scene was disturbed by hovering helicopters. A question was asked whether the wind generated by the helicopter could have distributed the blood present at the scene onto the undergarment. The bloodstain pattern analysis identified the stain on the bra (Figure 28-2) as a drip cast off (LVIS, gravitational, passive) that had spread on the edges of the underwire. Such a stain, which was round and well over the 6 mm diameter limit for a large

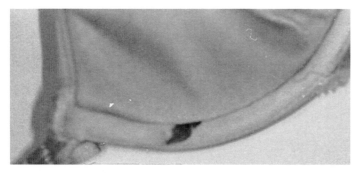

drop, resulted from the blood drop falling straight down onto a flattened bra. Some smearing was noted but was consistent with the curve of the underwire and absorbency of the bra material (Figure 28-3). The DNA type obtained was consistent with that obtained from the third male.

ADJUDICATION AND RESOLUTION

Charges of rape were dropped against the third man. Although the DNA results on the bra were consistent with his type, the evidence interpretation was that the bra was on the ground (seen in a photo), and that the defendant dripped blood on the bra in passing. It was noted by others present at the scene that the third suspect was walking around holding his hand where a cut was dripping blood. The type of stain on the bra was not consistent with acts of removing the garment, direct handling, or transfer pattern, which could be interpreted as initiation to the rape. In addition to interpretation of the results of the stain on the bra, witnesses at the scene, including the girls, stated the man was not involved in the rape. The prosecutor formally dropped the charges of rape after hearing the bloodstain pattern evidence in addition to the other information provided.

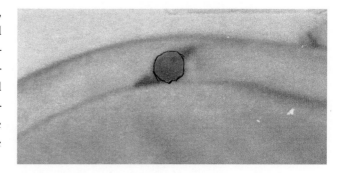

Figure 28-3
Outline of drip cast off bloodstain.

WHAT WE CAN LEARN FROM THIS CASE

When bloodstains are present, bloodstain pattern analysis and DNA typing can be powerful and effective when used in cooperation. Crimes may involve people having the right to be in an area, and those who may contribute blood due to injury. DNA may be instrumental in establishing possible contributors to a bloodstain, and bloodstain pattern evidence may help resolve how blood came to be where it was found.

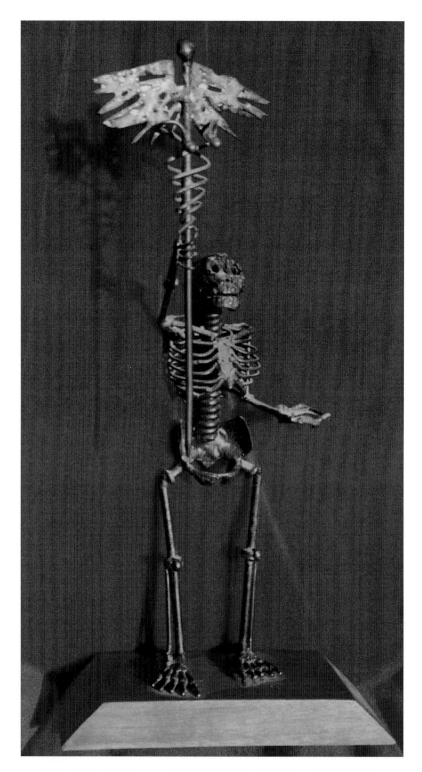

Figure 29-1   Weld scupture Medicine *by Lewis Lee Yount (1922–2000).*

CHAPTER 29

PATHOLOGY AND BLOODSTAIN PATTERN EVIDENCE: THE PREDOMINANT GOOD, OCCASIONAL BAD, AND RARE UGLY

THE PREDOMINANT GOOD

Although bloodstain pattern evidence began as a separate forensic discipline from pathology, the modern approach requires professional information regarding injury (i.e., probative blood sources). To prove a blood source was available for specific bloodstain patterns must involve medical professionals. In homicide investigation this will be a pathologist. Information provided by a good forensic trained pathologist can be essential to the interpretation of patterns and the whole reconstruction of a case.

Because the importance of bloodstain patterns is being conveyed to medical professionals, some are attending bloodstain pattern lectures and workshops, thus gaining understanding for their part in the evidence application.[1] Such training enhances the autopsy by emphasizing views of patterns on the victim before the body is cleaned and entered for internal examination. Stains on the body can be highly effective in court as evidence of "the victim figuratively pointing a finger at an accused." Recognition of the importance of exterior bloodstains also have led to coroners' protocols requiring care of clothing with ample photography before the body is placed in a bag for delivery to the morgue. Serious loss of evidence may occur from contamination of blood and body fluids when the body is transported.

In Chapter 21, the solution of the case and identity of the assailant who delivered the fatal blow resulted from the combination of a bloodstain pattern analyst and a trained forensic pathologist. The "pink" stain was noted on the wall of the victim's residence as well as on the jeans of the alleged accomplice. When the pathologist was asked "is it possible that CSF (cerebral spinal fluid) could be under enough pressure to exit explosively," the answer was not only that it could but that was what happened. The many blows struck to the victim's head would cause a substantial rise in CSF pressure and the one blow seen in the skull x-ray

---

[1]Kilely, Terrence F. (2001). *Forensic Evidence: Science and the Criminal Law.* CRC, Boca Raton, FL, 252.

showed that the area where CSF is produced was breached. The pathologist was qualified to state that this single blow was the one that would either immediately cause death or lead to coma and subsequent death. The bloodstain pattern analyst could not professionally state the cause of death, but could build upon the qualified statements from the forensic pathologist.

THE OCCASIONAL BAD

Hopefully infrequent but still an occurrence encountered in reviewing autopsy reports are situations where the autopsy is predominantly viewed as an anatomical report. Bloodstain patterns on the victim's body may be ignored, contaminated, or cleaned off before being properly photographed. Fortunately these kinds of errors can be covered with quality evidence photography and good collection procedures at the scene. Information from the autopsy is lost, however, and occasionally must be elicited during testimony in court. To do the latter (i.e., obtain what may be viewed as crucial information from the pathologist during trial) makes them appear less than professional in the eyes of the jury, and provides an inconvenience for both prosecution and defense.

In Chapter 5, case 1, the autopsy was conducted by a clinical pathologist who did not apply forensic training to that autopsy. Anatomically he followed procedure but omitted exterior comments. An apparent contradiction was brought out by the defense attorney during trial: blood was found in the victim's mouth, and yet the autopsy said no respiratory sources. During trial the defense attorney built a confusion regarding the evidence from the bloodstain patterns seen on the face by noting that respiratory projection could explain them. The mouth is commonly thought of as included in the respiratory system. Neither of these was in itself untrue, but they were made to appear as contradictions during cross-examination of the pathologist.

Although apparently bad for testimony the situation was recovered when the bloodstain pattern analyst pointed out that the flow running down the nose originated from the forehead, not the nose, and that blood from this flow later accumulated on the lips. Seepage into the mouth from the lips was on the surface of the victim's teeth, which qualified as blood in the mouth but not as part of respiration process. The pathologist's statements in both situations were thus reinforced and accepted by the jury.

THE RARE UGLY

Pathologists are as capable of error as anyone else, thus it is remarkable that finding such has been rare in the author's experience. Unfortunately they do occur. An example of dangerous error is when the pathologist accepts statements from law enforcement officers as facts to include in the autopsy. Such a situation may have occurred with the case in Chapter 13. The autopsy report described fewer wounds than were accounted for by the number of bullets the pathologist claimed were involved. The assailant readily confessed to shooting the victim and stated the number of shots she felt she fired. This exact same number was repeated at trial by the pathologist. No factual information was provided that corroborated and confirmed the number of shots. What appeared to have occurred is the law enforcement officers told the pathologist how many shots the accused said she fired, and that became the number stated in the pathologist's findings. Pathologists must be independent of suggestion by officers to provide just expertise in crime investigation.

In Chapter 25, another situation occurred that indicates a pathologist being influenced by law enforcement statements. The amount of coagulation that had to have occurred prior to water being added to the tub strongly suggested the victim was dead before water was present. The autopsy findings, however, listed heart failure and drowning, although no characteristics of drowning were stated. The only justification for the finding was that officers informed the pathologist that the victim's head was found in water.

Pathologists aren't the only physicians who may be intimidated by prosecutors and law enforcement officers. In Chapter 27, a neurosurgeon testified in court that the injury to the victim's head was specifically from being hit through a pillow by a golf club. The physician so testifying was at least the third physician to the victim at the second hospital, third admission (including ER), after the side of her skull was removed and the wound cleaned and dressed. Testimony was obviously based on what he was told to say, not upon medical facts known by him. He was a department chief of a large HMO hospital and would not normally work assault cases, could not have seen anything specifically indicating the exact weapon, which might not have been clear even to the primary emergency room staff. Sadly, defense attorneys may not feel confident in challenging statements made by medical doctors with illustrious qualifications in general, and wrongs may endure.

As the benefits from bloodstain pattern evidence becomes familiar to forensic pathologists, medical professionals, and the lawyers who dispose them, it will be less likely that bad and ugly situations will be accepted without challenge. Instead teamwork will be more common as the next case example of the predominant good illustrates.

CORRELATING PATHOLOGY AND BLOODSTAIN
PATTERN EVIDENCE

by G. Michele Yezzo, Office of the Ohio Attorney General,
Bureau of Criminal Identification and Investigation, London,
Ohio; and Diane Scala-Barnett, M.D., D.A.B.F.P., Forensic
Pathologist, and Deputy Coroner, Lucas Co., Ohio.

BACKGROUND

On June 26, 2000, Logan County, Ohio Sheriff's Office was provided with information about a possible homicide. Based on witness statements, the Sheriff's
department initiated an investigation.

After obtaining a warrant, a search was conducted on the subject's residence.
During the search bloodstains were noted, including those on surfaces in a bathroom. Information provided by one of the witnesses suggested that the body
would be found in a rural area of a neighboring county. Investigators from that
jurisdiction were notified of the situation, and asked to assist in the search.

The body of a woman in an advanced state of postmortem decomposition,
later identified as the victim, was found in a ditch in the neighboring county.

BLOOD SOURCE IDENTIFICATION

The deceased died of blunt force injuries of the neck and chest due to
beating.

BLOODSTAIN PATTERNS IDENTIFIED

The bathroom of the deceased's residence was in a state of disarray as if it were
being remodeled. However, an impact pattern was demonstrated on the front of a
built-in vanity (Figure 29-2), and another impact pattern was demonstrated on the
adjacent wall (wallpaper partially removed prior to the bloodstains being deposited Figure 29-3). Spattered stains were visible on a cardboard box next to the wall.
Due to apparent warping of one of the drawers it did not close flush to the face
of the vanity. Spatters, identified as part of the impact pattern on the vanity, were
demonstrated on the underside of this drawer (Figure 29-4). Contact stains and
isolated drips were noted on the plywood surface of the bathroom floor.

AUTOPSY FINDINGS

An autopsy was performed on June 28, 2000. Identification of the individual was
made by comparison of premortem and postmortem dental x-rays. Although

Figure 29-2
Impact pattern on the front of the built-in vanity.

the body was in an advanced state of postmortem decomposition with insect feeding, traumatic injuries were still discernible and were confined to the head and neck. There was a fracture of the larynx at the cricoid cartilage (Figure 29-5), and rupture of the body/greater cornu synchondrosis of the hyoid bone, bilateral rib fractures, and a fracture of the cervical spine (C6) with epidural hemorrhage of the spinal cord in the cervical and thoracic regions. At the point of fracture, there was a laceration of the spinal nerve root. It was the opinion of the forensic pathologist who performed the autopsy, that the mechanism of the injury was a kicking or stomping action as the forces were concentrated or localized to one area. Because of the condition of the body, strangulation could not be ruled out.

Figure 29-3
Impact pattern on the wall (wallpaper partially removed prior to the bloodstains being deposited).

OBJECTIVE APPROACH TO INTERPRETATION

Although there were irregularities in the surface of the vanity face, it was possible to correlate areas of convergence of the impact patterns on surfaces of the vanity and the wall of the bathroom to ascertain that the blood source was below the level of the second drawer from the bottom, indicating that the blood source had been very close to the floor at the time of the impact.

Figure 29-4

Spattered stain, part of impact pattern on the underside of the vanity drawer.

Figure 29-5

The fracture of the cricoid cartilage in the larynx.

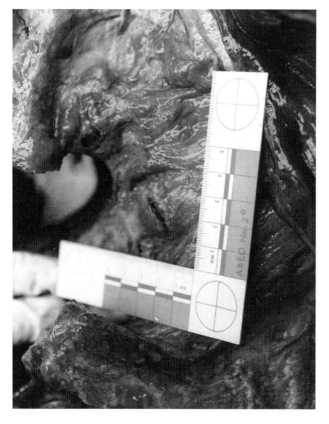

RESOLUTION, ADJUDICATION, AND FINAL DISCUSSION

Testimony was given by both the bloodstain pattern analyst and pathologist on this case, which helped to provide a clear picture to the court and jury of the violence of the incident. The defendant was found guilty of murder, felonious assault, and abuse of a corpse.

WHAT WE CAN LEARN FROM THIS CASE

Independent analysis of the physical evidence from the scene of the incident (residence) and finding from the autopsy may provide corroborative evidence. In this case, although the body of the victim was badly decomposed, the cause of death could be determined as a result of a forensic autopsy. Correlating the available blood sources (injuries) with the impact patterns provided valuable information.

Many thanks to the family of deceased, Vicky Grubb (Figure 29-6) for authorizing the use of material from this case. Seldom do forensic scientists see appreciation for their work from the victim's survivors. Gratitude is also extended to Logan County, Ohio Sheriff's Office Chief of Detectives Jeff Cooper and Evidence Technician Phil Bailey for providing investigative information.

Figure 29-6

Photograph of Vicky Grubb and family, survivors.

Figure 30-1   Lady Justice carries a sword dripping blood.

CHAPTER 30

BLOODSTAIN PATTERN EVIDENCE AND THE LAW

INTRODUCTION: A SCIENTIST'S OPINION OF LAW

DICHOTOMY OF VIEWPOINTS

Bloodstain pattern evidence (BPE) often is approached in a legal manner rather than as scientific evidence. In order to understand how this can influence the application in bloodstain pattern analysis (BPA), we must first review the dichotomy between science and law, which is forensic science. In criminal law there are frequent debates[1] of opposing viewpoints, and currently there is a division of opinions in BPA regarding terminology and definition of science principles. Although some methodologies in BPA were probably in use thousands of years before fields in science and law existed, technical progress in BPA has been slow and occasionally based on politics rather than science principles. The first experts were law enforcement officers who, as *shire reeves*, were appointed by the king or other noble to judge and manage public affairs. It is not surprising then that until scientists entered forensics, there was nothing recorded that suggested debate or disagreement regarding how the evidence would be applied in hunting, tracking, and predominant property crime investigation.

The differing opinions about bloodstain pattern evidence emerged sometime around the 1950s and 1960s. During this period, emphasis was shifting from crime labs staffed predominantly with police officers who progressed to that duty station, to scientists from academic backgrounds, most with no police training. Without acknowledging differences, the groups from law enforcement and those with science backgrounds began to disagree on terminology and some training techniques in BPE. Unfortunately, there were also those who attempted to eliminate opposition, and figuratively write a single viewpoint in stone, censoring others not accepting their perspective. The attitude that one view is right and another view is wrong reflects a legal or political approach more than a scientific one.

---

[1]Reynolds, Quentin. (1950). *Courtroom.* Farrar, Straus & Co., New York. *Quotes from Samuel S. Liebowwitz.*

The dichotomy of BPE can be described as the difference between those favoring scientific definition, and advocates of practical explanations, who might lack scientific principles. Forensic scientists, educators, legal scholars, and researchers need to understand the true scientific nature of events, procedures, and analysis in order to evaluate accuracy of BPE and develop new procedures and techniques. On the other hand law enforcement investigators, trial lawyers, basic crime scene processing trainers, and the public (from which jury pools are derived) require only information regarding applied and practical aspects of the evidence sufficient to answer specific questions like, for example, whether bloodstains on the accused clothing place them at the scene when the victim was beaten, or whether the bloodstains are consistent with the victim trying to defend him- or herself during the assault.

The same evidence needs to be viewed from various focuses by different parts of the investigative process. Debating who is right and who is wrong is futile. The focus should be on what is right with regard to the evidence and what benefits the details can provide.

Lawyers are the key to the income and professional success of individuals wishing to work in this field. Unfortunately there is no nonforensic alternative uses for BPE such as exists for DNA (fields of genetics), trace (manufacture of articles such as fibers, glass, tools), and toxicology (various application in medicine and chemistry). There is, therefore, no economic source to fund employment or research in BPE outside the criminal, and occasionally civil, justice systems. The possible resolution of the dichotomy is that lawyers become aware of different levels of terminology, not to reject or favor one, and that complete knowledge of the technical basis for the science is unnecessary for explaining practical applications.

DIFFERENCES BETWEEN LAW AND SCIENCE

It should be understood that the logics of law and science are dissimilar, and that these differences are carried further in BPE as a science. At least four fundamental differences in logic may influence communication between lawyers and bloodstain pattern analysts:

- Bivariant versus multivariant systems
- Words versus things
- Start with problem versus start with the parts of the problem
- Frequent change versus slow process of gradual change

Law tends to be bivariant, requiring choices between two alternatives such as evidence being viewed as a match or no match; testimony is true or false; an accused is guilty or not guilty; answers yes or no; or is consistent or inconsistent with. Science is multivariant, and is better viewed as a formula rather than as a choice between opposites:

$$A + B = C - D$$

This is A plus B equals C minus D; however, not all As and all Bs will yield all Cs, which may or may not need to have some but not all Ds subtracted. Accurate scientific answers sometimes take the form that: the transfer pattern matches in some parts but not in others, the statement would be true if certain conditions exist but false if other conditions are present, and the evidence suggests involvement or lack of involvement in some aspects but does not indicate either guilt or lack of guilt. There are questions for science where neither a yes nor a no would be a wholly true answer. On the other hand our adversarial form of justice frequently imposes bivariant legal requirements. If time permits, and admittedly it often does not in casework, a thorough examination of various interpretations can lead to a conclusion that may completely or partially encompass both sides of apparently divergent opinions. A point with BPE that is seldom appreciated in law is that it may be more productive to reconcile different interpretations than to attempt to discredit either.

Law is a matter of words, which are two-dimensional in context. Definitions may be questioned in law. Arguments over semantics are common at meetings of bloodstain pattern specialists. Different dictionaries, however, may include varying definitions for some of the same words for different disciplines: medical, engineering, law, and English usage have versions defining terms differently for specialists. Bloodstain pattern evidence requires the same allowance for different semantics of biomechanics, pathology, hematology, blood rheology, law enforcement, academic research, and criminalistics.

Science is a matter of things, rather than just words. Things are predominantly three-dimensional in context. An example is in Microbiology and Infectious Disease conferences, where doctors may discuss *typhoid bacillus*, *typhi bacterium*, Salmonella typhosis, and *Salmonella typhi*. These all may be the same microbe, but the importance in discussion is the focus toward a disease, its causes and treatment, not that everyone use the same bacterial name. In contrast, committees of bloodstain pattern analysts may become stuck in discussions regarding whether or not members can use particular terms and still be regarded as qualified experts. The semantics take on an importance, yet science principles are to be accepted without question.

One of the major differences between law and science in handling bloodstain pattern evidence is that law starts with a problem, such as a crime, and then searches for acceptable parts: precedents, motions, statutory law, court rules of order, real and circumstantial evidence to fit a prosecution or defense viewpoint of the problem. Science on the other hand starts with parts and then sequences them to define the problem. For example, DNA identifies the victim, trace evidence connects the accused to the crime scene, and BPE shows the dynamics that occurred between the victim and the accused (or the clothes worn by the accused) in an assault. The parts used together then identify a person to be charged with a crime. When discussing lab results, scientists and lawyers may not be applying the same logic. Dissatisfied lawyers may subsequently ignore their normal science contacts and involve others who may be willing to make less supportable statements. Bloodstain pattern evidence then is used in a strictly legal approach, not as a science.

A dilemma for lawyers and scientists is the difference between the rates of change for science compared to law. Law tends to work slowly. The rules that govern civilization need careful consideration before change. Social instability would result were laws frequently revised. Therefore each process must be studied for probable effects and acceptability with other existing laws before alteration. Law changes gradually over long periods of time. Science, especially since the 1950s, has changed so rapidly that technology regarding something like computers from a year ago may be completely obsolete now. The legal system acceptance rate may be viewed as too gradual by scientists, with science principles lagging too far behind in court acceptance.

REASONS TO HIRE AN EXPERT

A traditional approach for attorneys to bloodstain patterns in trial is to focus on admissibility of the evidence or the expert. This trial strategy has been shown to be inadequate in many situations. The evidence is commonly admitted,[2] thus attempting to show it isn't a science is contradictory to the better approach of submitting interpretations from their own expert. The infamous O. J. Simpson trial illustrated the dangers of not putting on all your evidence even when you have DNA.[3] Prosecutors and defense attorneys should give serious thought to employing their own experts where bloodstains may provide information regarding a crime. The benefit of hiring an expert can be considered from three perspectives:

- Prosecution
- Defense
- By Direction of the Court

The process of using an expert is remarkably similar from the prosecutor's or defense's viewpoint.

---

[2]Giannelli, Paul C. and Imwinkelried, Edward J. (1999). *Scientific Evidence, 3e,* Volume 2, 24-13(B). Michie Company, Charlottesville, Virginia, 498.
[3]Bugliosi, Vincent. (1996). *Outrage, the Five Reasons Why O.J. Simpson Got Away with Murder.* W.W. Norton, New York, 218–219.

COMMENTS REGARDING EXPERT WITNESSES IN BPE

Several years ago at an AAFS annual meeting, the author was in the bar area jotting down notes, when a man sat down at the table. He looked searchingly as if to understand some alien creature. When the author looked up and smiled, the man asked, "but you are deaf, how can you do this when you can't hear?" *This* apparently referred to forensic science. My answer was, "I have a brain and eyes, and I still use them."[4] Later that day some groups were in an elevator going to dinner, when I related the encounter to friends. Dr. James Ferris, former IAFS President, overheard the remark and commented, "I know plenty of people who have neither and are still testifying in court." Although this was said in humor, it is a common concern for the legal establishment. Forensic scientists express the opinion that attorneys should weed out those individuals who are incompetent, insufficiently qualified, or dishonest. This may be a contradiction in practice.

The adversary form of justice, where the defense motto is, it is better to allow a guilty person to go free than to incarcerate an innocent one, opens the door to abuse, yet anything less than a rigorous defense can create other problems, leading to further injustice. Both prosecution and defense may employ whatever means available, and whoever can reasonably deliver those means, to defend or convict the accused.[5] It is unrealistic to expect an attorney to ignore someone who tells him or her what the attorney wants to believe. The best way to deal with the situation is not censorship of individuals, but rather for opposing counsel to use better experts relying on better science principles.

SPECIFIC BENEFITS FROM BLOODSTAIN PATTERN EVIDENCE FOR DEFENSE ATTORNEYS

Bloodstain pattern evidence can provide the following benefits:

1. It may show that the defendant is not guilty *as charged.*
2. It may demonstrate that there is more evidence than disclosed (i.e., worst case scenario).
3. It may encourage the client to drop insupportable claims of innocence and accept a negotiated plea.

The accused and witnesses involved in some cases may distort the truth even when not germane to the issues. Detectives are trained to regard witnesses and suspects lying as indications of guilt. This ignores two possibilities: some people lie even when unnecessary, and some have poor memories, especially when at

---

[4]This remark was adapted from a wiser woman who explained how she could be a mother and a professional. Her reply was something like, "I have a uterus and a brain and I can use both."
[5]Reynolds, *Courtroom,* 24–25.

the time of the event they were unaware they needed to remember accurately. Having a bloodstain pattern analyst reconstruct the events of a blood distributing crime before taking a client through the steps of an assault can prepare an attorney to critically interview his or her client. A possible advantage could also be in establishing that a lesser crime was committed than that for which the client is being charged.[6]

The author's personal experience with cases is that in many situations the client left more evidence than law enforcement investigators had time and staffing to uncover. Preparing the defense for the worst case scenario gave them an advantage to plan trial strategy and to decide on a better negotiated plea for their client.

Many accused individuals will initially claim versions of the events that are totally unrealistic. If faced with incontrovertible scientific facts early in the investigation, many are willing to accept negotiated pleas rather than continue to trial. This expedites resolution and benefits both prosecution and defense. For an equal playing field, each side should employ their own bloodstain pattern expert.

The use of court-appointed experts would benefit both sides by putting the expense and loyalty of an expert on the court's account. Unfortunately this is not the usual means. *Amicus curiae*, literally friends of the court, usually provide information at their own expense on behalf of the accused or convicted. The most famous *amicus* brief for BPE was following the Sam Sheppard conviction in which Dr. Paul Kirk illustrated the importance of bloodstain patterns in reconstructing the events of a murder for the defense seeking a retrial.[7]

WHERE ARE EXPERTS FOUND?

There are several sources of experts available today: law enforcement identification and homicide officers; criminalists; private investigators; college professors; clinical laboratory scientists; and law enforcement officers from duty stations other than homicide and identification. Those claiming to be experts usually have one or more of four types of training:

- Experience of viewing many bloody crime scenes
- Reading material; several textbooks and many articles available
- Less than 40-hour short programs or lectures within other programs
- One or more 40-hour programs

---

[6]*Ibid*, 26 f.
[7]Kirk, Paul Leland. (1955). Affidavit on behalf of Samuel H. Sheppard Court of Common Pleas, No. 64571, State of Ohio, Cuyahoga County.

Self-training is generally inadequate unless the individual has had access to considerable experimentation in bloodstain patterns using real human blood. Law enforcement experts tend to emphasize experience over academics. The number of crime scenes attended does not relate to individual ability to identify and interpret what they have seen until after training. Also some participants with one to eight hours of instruction included within Death Investigation, Introduction to Investigations, and Advanced Investigations programs regard the lectures as sufficient to apply BPA and testify regarding interpretations. Introduction lectures create interest in further training and illustrate the benefits of a full 40-hour program, but do not qualify participants as experts.

Forty-hour programs are necessary to begin expertise in the field. Experience does not begin to qualify an individual until after they have had proper training. For that matter, not all 40-hour courses provide confidence and understanding that assures participants can apply methodologies to actual cases. The best programs include a mock crime scene. As discussed in Chapter 4, some mathematical applications have logic flaws that should not disqualify the evidence but do merit recognition and confirmation before drawing conclusions. In order to assure that students have learned correctly and appreciate limitations of the evidence, a realistic mock crime scene is essential. Some agencies also use an apprentice program, which should improve the standards of experts, if those guiding the apprentice use scientifically sound approaches.

QUESTIONS AND ANSWERS WITH EDWARD IMWINKELRIED

Professor Edward J. Imwinkelried, University of California Law School at Davis, has generously provided answers to questions posed by the author since 2002. This is especially beneficial since Prof. Imwinkelried consulted in the preparation of the original Daubert brief[8] for the plaintiff. The following are some of the questions asked and answered.

GENERAL QUESTIONS

Q. Why do attorneys have such faith in bloodstain pattern evidence even if the logic is flawed?

A. In part, many attorneys have such faith in bloodstain analysis because of *Frye*.[9] Prosecutors successfully demonstrated that the technique was widely accepted among trace evidence experts; and until Daubert, the bar gener-

---

[8]*Daubert v. Merrell Dow Pharmaceuticals, Inc.,* 509 U.S. 579 (1993).
[9]*Frye v. United States,* 293 F. 1013 (D.C.Cir. 1923).

ally did not appreciate the importance of critically evaluating the underlying research which supposedly validates a theory or technique. Even in *Frye* jurisdictions, you can use Daubert style empirical reliability arguments to attack the weight of the opponent's testimony.

Q. Aren't bloodstain patterns physical evidence? Why are they referred to as circumstantial? I'd expect blood from a victim would be a direct link to the crime, not one of circumstance.

A. There are two relevant distinctions. One distinction is between substantive evidence and credibility evidence. Substantive evidence is admitted as proof of historical facts in issues on the merits of the case. Credibility evidence is admitted solely for whatever light it sheds on a witness's credibility. BPE is substantive evidence.

The second distinction is between direct and circumstantial evidence. Eyewitness testimony illustrates direct evidence. If the question is who committed the crime and the witness testifies that she saw D commit the crime, her testimony is direct evidence; and the only question for the trier of fact is whether to choose to believe the witness. In contrast, circumstantial evidence is simply proof of a link in a chain of reasoning leading to an inference as to an ultimate fact in issue. Most scientific evidence is circumstantial in character. BPE is circumstantial evidence.

Q. Are juries more likely to convict in a bloody crime because they fear a greater evil would be freeing a violent offender? This would be, in especially violent and bloody assaults, to feel it is better to incarcerate an innocent person that to let a guilty person go free.

A. Most jurors take their job and the judge's instructions seriously. Consequently, at a conscious level the overwhelming majority do not make a decision that it is better to imprison a bad person even though he or she may be innocent of the charged crime. Rather, the issue is how much weight they attach to the bloodstain evidence. Critics would argue that they erroneously convict because they ascribe undue weight to the evidence. At a subconscious level, they might be especially tempted to do so when the crime was a horrific one.

Q. In my experience, some attorneys demand reports that contain specific wording that is either imprecise or inaccurate from a scientific perspective. How does an expert tactfully tell the attorney that the question's wording is deficient?

A. On the one hand, when the expert is on the witness stand, he or she can only respond to the questions asked—the expert cannot simply reword the question and answer the question he or she wishes the attorney had asked. However, even on the witness stand, if the witness honestly believes that the wording of the question is vague or misleading, the witness can bluntly say so. Hopefully, that will force the attorney to rephrase the question. On the other hand, when the expert is merely consulting with an attorney, the expert can feel free to say, "The question you asked uses the expression, '*low velocity impact spatter.*' I think that you're using that term in an imprecise way. Allow me to explain …."

ABOUT DAUBERT

Q. My experience has been predominantly with *Kelley/Frye*.[10] Can you give me an idea how Daubert compares?[11]

A. *Daubert* and *Kelly-Frye* differ in several significant respects, including the following two:

1. THE NATURE OF THE STANDARD. *Daubert* requires a showing that the expert properly used scientific methodology to empirically validate the theory. In contrast, *Kelly-Frye* requires a showing that the expert's theory or technique is generally accepted within the relevant specialty fields.

2. THE SCOPE OF THE TEST. As modified in Kumho (1999),[12] the Daubert reliability requirement applies across the board to all types of expertise. In many Kelly-Frye jurisdictions, the test is limited to novel hard scientific theories and techniques. Consequently, traditional techniques, soft science, and non-scientific expertise are generally exempt. *Daubert* makes it relatively clear that in order to qualify as "scientific … knowledge," evidence must rest on methodologies acceptable to scientists—not police officers. Moreover, in *Kumho* in 1999, the Court stated that it wanted to ensure that the experts "employ in the courtroom the same level of intellectual rigor that characterizes the practice of an expert in the relevant field."

Q. You've mentioned "novel evidence" a few times in regard to how new discoveries can be submitted for admissibility evaluation. What if terminology with definitions was established as acceptable in court at an earlier time, then changed to a completely different meaning later? Given that the change in terms would include changing methodology in training and applications, would returning to the original terminology and definitions constitute novel evidence? Use the term low velocity impact spatter as an example.

A. If you change the definition of low velocity impact spatter in a science proposition, you change the proposition. Hence, whether the governing law is *Frye* or *Daubert*, the prior admissibility of testimony based on proposition #1 neither dictates nor even justifies the admissibility of testimony based on proposition #2.

---

[10]*People v. Kelly* 17 Cal 3rd 24 (1976), 549 P. 2nd 1240.
[11]*Daubert.*
[12]*Kumho Tire Co., Ltd. v. Carmichael,* 526 U.S. 137 (1999).

Q. What if we want to return to an earlier definition of the term?

A. You simply want to identify the underlying scientific proposition. In the final analysis, neither *Frye* nor *Daubert* courts accept testimony based on labels such as low velocity impact spatter. Rather, they accept or reject a certain proposition as a permissible basis for expert testimony. The expert should tell the attorney—in the hope that the attorney will tell the judge—to look past the label and identify the underlying scientific proposition.

QUALIFICATIONS OF EXPERTISE

Q. How can someone with a better science background protect themselves and their testimony when appearing opposite someone with less academic science but more investigative experience?

A. Given that state of the law (*supra*), you might convey the following suggestions to the attorney putting you on the witness stand:

1. You might tell him or her that you anticipate this line of cross-examination and suggest that he or she move *in limine* to preclude the line on the ground that it is irrelevant, since it does not relate to compliance with scientific standards. After all you propose testifying solely about the scientific soundness of the BPE in the case. Even if the judge does not grant the motion, litigating the motion gives the attorney an opportunity to educate the judge about the issue.

2. You should be primed for a redirect examination to clarify that very point. If the cross-examiner is good, he or she will ask tightly phrased, leading questions that might not give you the opportunity to volunteer the explanation on cross. The attorney needs to know that you have an explanation and be prepared to allow you to present the explanation on redirect after the cross.

An example of the benefits of teaching the court occurred during the Alexander Lindsay second inquiry in Sydney, Australia. The crime, a blunt force assault on Pamela MacLeod Lindsay (see Chapter 6), produced bloodstain patterns classified as arterial spurting, swing and cessation cast offs, and blunt force impacts. On the eighth day of testimony, opposing counsel asked if the spatters were all medium velocity impact spatters. The witness was ready for this question and prepared to answer, "all the patterns would be in the medium velocity range but not necessarily the results of impact." However, before the witness could answer, the judge asked for clarification from the Crown barrister, "do you mean to ask if the bloodstains were arterial, impact, or cast off?" The judge had been taught the difference between dynamics.

The importance of this short exchange was that the evidence itself was the focus. If the judge and jury understand how the interpretation (analysis) is made, the resume or lack thereof of the expert is no longer important.

A SCIENTIST'S WISH LIST FOR THE FUTURE OF BPE

I have had the good fortune to work with conscientious and impressive attorneys on both prosecution and defense sides. In my experience, some practices would improve expert-attorney communication. It is wished:

1. Attorneys appreciate the importance of the evidence early in the investigation. Too often it is brought in during or immediately before trial when strategy is already formed. The information from BPE may change conclusions and suggest a different strategy. As a practical matter, it can be too late to change direction after trial begins.

2. Judges realize that spatter experiments do not produce picture matches to crime scene evidence. The reason to call an expert is to identify and interpret the meaning of patterns found at the crime. Some patterns are composed of spots called spatters. It is unrealistic to expect the trier of fact to conclude what is a match between experimentally produced patterns and the crime evidence. Fingerprints and contact transfer patterns are pattern match, and the logic for stating a match can be shown. Spatter patterns are based on too many variables to match a designed experiment. Simple sizes can be reproduced in several ways that do not prove a match. Without the benefit of sound scientific insight, the trier of fact is not qualified to recognize what is a pattern match. That is why an expert was called in the first place.

3. Experts lacking science background in bloodstains realize they should not attempt to elaborate for the court on the technical background of blood behavior. Practical information within their experience and training may be supplied to the trier of fact without going beyond qualifications. If they are unqualified to explain the scientific underpinnings of BPE, they should not be pressured into overstepping their limits.

4. Examining trial attorneys critically question experts as to how they arrive at the identification of pattern classifications, especially velocity terms, rather than automatically accept the expert's opinion. This approach would help discourage the use of BPE as "instant evidence," i.e., claim the VIS patterns specifically identify the crime and that the accused committed the assault. In many cases the pattern could have resulted from alternative events not necessarily criminal.

5. Attorneys, judges, and juries be aware that the discipline is a lot more than just "blood spatters." It is a complex, growing field of physical evidence that can be crucial to the adjudication of crime. This is why the modern terms of BPE for the evidence and BPA for the expert application is more encompassing.

EXPERIENCE WITH TRAINING IN BLOODSTAIN PATTERN EVIDENCE

Figure 31-1   A Tradition in Bloodstain Pattern Evidence workshop has included fun as well as education. Here a class gathers around a drying sequence exercise.

CHAPTER 31

PREPARATIONS FOR BLOODSTAIN PATTERN WORKSHOPS

APPROACH TO BLOODSTAIN PATTERN TRAINING

A primary caution for training programs in bloodstain pattern evidence (BPE) is in teaching investigators how to analyze bloodstain patterns without emphasizing pattern match approaches. Many experienced instructors will shake their heads at this, feeling that pattern match is an essential to training. This, in a specific way, is also true. So how can something be wrong and essential together? The key is in who decides what is or is not a pattern match, and how they define such. Experienced instructors and investigators see a lot more than just a bunch of red (brown or black) spots on a white cardboard sheet. They have learned subconsciously, if not consciously, to integrate qualifying characteristics so that in their minds the patterns seen match similar arrangements representing dynamics previously encountered. Characteristics that may be noted are:

- The whole arrangement of spots representing one dynamic event
- Directions of travel of individual bloodstains
- Recognition of spatters associated with overlapping patterns from subsequent events
- Understanding of shapes for groups of spots generally associated with specific events
- Position and placement of spots with respect to each other and the whole
- Distribution of sizes and importance of size ranges within the whole
- How other evidence confirms or refutes identification of pattern categories

Students new to the evidence see only a bunch of red spots because they have yet to learn how to integrate details into an identification scheme. Implying from the beginning of a workshop that bloodstain patterns are pattern match is dangerous because it encourages participants to try to memorize what they see in the lab sessions, rather than learn why and how things resulted as they did. Since students have not yet learned what are reliable qualifications for pattern match they tend to focus on tangible criteria such as the mere presence or absence of spots, or the size of individual spots making up patterns, while ignoring arrangements, alignment, and distribution of size ranges within an organized pattern. It has long been established that individual spatter size involves too much overlap to use as a sole identification between different events such as impact, cast off, and arterial breach, and also within noncriminal acts such as respiratory distribution, splash, and blood into blood.

Another problem with focusing on size alone is that during a workshop, patterns are isolated and created in a uniform manner. Sizes are more predictable than they will be at the scenes of violence. Simplified exercises do not show the multitude of interchanging variables that can be encountered in casework. This leads to newly trained individuals falling back on simplified principles in attempts to memorize.

It is, therefore, essential to teach investigators new to this form of evidence how criteria are included in the identification of pattern categories. Exercises must resemble what the participant will encounter at actual crime scenes, if the viewed patterns are to provide criteria reflected in actual casework later. The benefit of this approach is that graduates of programs will be able to identify events regardless of the context, surface textures, crime scene arrangement, or source of blood. With experience, students will learn how to associate certain patterns as belonging with specific dynamics whenever and wherever they are found.

The following material is a culmination of 29 complete 40-hour programs and numerous short advanced and specialty programs with dozens of lectures and scientific paper presentations. Some exercise concepts worked extremely well, some needed modifications, and some failed in class settings. The sum of these experiences is presented here to aid others designing programs so that they need not waste time repeating our bad experiences. Further modifications will no doubt be better and continue to increase the knowledge and benefits from BPE.

BLOOD PRODUCTS AND SUBSTITUTES AND BIOLOGICAL HAZARDS CONSIDERATIONS

BLOOD PRODUCTS, SUBSTITUTES, AND BIOHAZARD CONSIDERATIONS

It should be a prerequisite of any Bloodstain Pattern workshop, where actual experiments with blood (any blood, including animal and participant volunteer) are to be conducted, that all attendees have completed a biohazard lecture program. These are available at all law enforcement agencies known to the author, are usually mandatory for all sworn officers, and provided for a number of professions and trades beside law enforcement.

When training first began, blood-borne pathogens were of less concern than whether or not one could include firearms exercises. In 1986 the attitude toward biological hazards shifted completely toward concern for exposure to a proven fatal microbial disease acquired from contact with blood products. Since then HIV has been proven difficult to acquire, but hepatitis, which is easy to acquire, is becoming more of a concern than previously during classes with law enforcement officers.

The initial fear of HIV and AIDS hit law enforcement especially hard. The "job" requires considerable danger and concern for safety from visible threats. Dealing with an invisible, potentially fatal, disease created something resembling panic. Participants in California bloodstain pattern workshops in the late 1980s sometimes insisted on gowning up to look much like astronauts on the moon. Although such extreme measures were not necessary, the outcome was beneficial in an indirect way. The first 40-hour workshop held in Sacramento identified much of the class as having had hepatitis. Hepatitis A and B have been known for several decades, with the present addition of C as another blood-borne virus. Hepatitis seldom kills but frequently creates permanent health damage to the liver and the ability to transmit the virus to others. The fear of AIDS led to crime scene procedures that have reduced exposure to hepatitis. Workshops held in the last few years have had no participants reporting positive to hepatitis testing. Instructors do well to emphasize that precautions for AIDS are also beneficial to prevent other diseases.

When AIDS and the HIV virus were identified and described to trainers in law enforcement, the first thoughts were to eliminate blood from the workshops. Some agencies still focus on this and attempt to provide training without using the essential substance of bloodstain patterns. In colleges and academics, this is a good idea and attempts to find agreeable substitutes will continue. With law enforcement, however, the job situation involves many hazards that are covered by the initial oath from police academy graduation. These say, in effect, that if a person is accepting the job of being a law enforcement officer they are also accepting all the dangers inherent with the profession. This means a sworn officer cannot sue the department for injuries in traffic accidents, gun battles, or acquiring a disease while collecting evidence from a drug bust.

If an officer is to accept the hazards of the job, however, he or she must be properly trained in what those hazards actually are. The best way to deal with biological hazards presented by bloody crime scenes is to show students how to act around this evidence while also learning the investigative leads potential. Therefore the best material to use in police and crime lab staff training is real human blood that has been tested for all known blood-borne pathogens. Blood that is suitable within the United States, United Kingdom, Canada, Australia, New Zealand, and other countries with strict guidelines regarded as acceptable for transfusion can be considered safe for Bloodstain Pattern workshops. The attitude toward exercises, however, is for the students to treat all blood as if it is known HIV and hepatitis carrying. Instructors should observe participants doing things that would jeopardize their health (i.e., putting pencils or anything else in their mouth, using cell phones while still wearing protective gloves, and eating, drinking, or smoking without removing protective clothing and washing hands).

Some instructors feel safer using animal blood. Pig blood has been shown to be closest to human, but pigs carry viruses that are dangerous to humans but do not show up in the animals (i.e., the pig donors seem healthy but their blood may be hazardous to humans). The same can be said for horse, cow, and sheep blood. It should be remembered that HIV came from an animal before it infected humans. Media coverage of bird flu can provide some idea of the problems with using blood from animals in an environment where humans become exposed. Most serious is the false sense of safety. Law enforcement people are aware of the hazards of AIDS and some treat human blood with more respect than they treat firearms. Animal blood may find a more relaxed approach, which does not benefit training. If any kind of blood product is used in a workshop, it should be treated as hazardous for practice with body fluids as well as true safety precautions. Practice makes perfect.

For eliminating blood completely, latex-based paints, glycerol and food coloring, and dyed milk can all be used to show such features as direction of travel and whole pattern shape grouping of spots. For basic exercises in an unprotected situation such as secondary or middle school curriculum, these would be quite acceptable. If a substance is to be dyed with red color to simulate blood, a suggestion is to use cake paste type dyes, which do not dilute. These are available from bakery supply stores. For those whose job it will be to interpret bloodstain patterns at crime scenes, nothing works as well as real blood. Some exercises require blood to produce specific patterns. Transfers in blood are unique because of the red particles more or less suspended within a clear or slightly opaque beige liquid. Direction of movement, aspiration, settling, and sequential transfers require blood substance to be fully appreciated for the information provided with the evidence.

Volume (pooling) experiments require blood coagulation to duplicate drying times and lack of hygroscopic ("water loving" or taking up moisture from the air) behavior. The traditional procedure of using outdated human blood for volume stains is greatly affected by anticoagulants (calcium complex systems), which results in day-to-day weight shifts due to take up of atmospheric moisture and subsequence evaporation. Adding excess powdered calcium (black board chalk dust) can force some degree of coagulation and help decrease weight variance (note results in Chapter 25).

Some workshops have drawn fresh blood from volunteer participants. This is not recommended because the volunteers are not known to be disease-free, and hepatitis may be common among older crime scene investigators. Individuals may also not be aware they were infected with hepatitis. Biohazard lectures seem to be improving behavior around crime scenes, so newer officers may be hepatitis-free. If class volunteers must be used, select those who have blood bank donor cards, meaning their blood has been tested and found disease-free for donations. The anticoagulant of choice is EDTA (ethylene diamine tetra-acetic acid, lavender-topped Vacutainer® brand tubes). Other anticoagulants may be insufficient to prevent clotting (oxalate, black tops) and sodium fluoride (gray-tops, blood alcohol tubes), or to avoid hemolysis upon lengthy standing (citrate, blue tops), or just be expensive (heparin, green tops).

Mixing plasma units from outdated fresh frozen plasma (FFP) units with packed red cells to achieve a whole blood ratio of red cells to plasma is not advised from the author's experience. Even being cautious to match serotypes, type A plasma to type A cells, or at least compatible combinations such as O cells and AB plasma, etc., still involve micro-clotting when used in experiments. Such mixtures used in the artificial artery clogged the system to the extent that it was not operable until after a thorough bleach cleaning. Experiments without instrumentation were possible but small clumps of loose clot material were occasionally seen on the targets. Explaining microclotting to participants was more confusing than just using packed red cells in the first place. For priming mechanical devices, diluting packed red cell units slightly with saline was better than plasma additions.

HANDLING OF BLOOD

Although human blood from outdated transfusion pouches is the best choice for workshops, it is essential that it be from a certified American Association of Blood Banks member. Some blood suppliers will sell units to instructors putting on experiments or workshops. It is good practice and smooths communication to have handy a copy of the safety procedures you will be using when you purchase body fluids. It is also important to arrange discarding of the pouch and left-over product after use. Most garbage dumps do not permit discarding of biological hazard waste in regular trash bins.

A note regarding discarding of bloodstained materials after experiments or workshops is that any public trash containers may be examined by homeless and drug addict populations. Therefore trash containing syringes, needles, and bloodstained items should be discarded with thought to prevent providing sources of disease (i.e., never in public receptacles). Although garbage pickers may not be a concern, the civilian population they encounter will follow as a problem. Discard is best arranged in advance with a blood bank, local hospital, or coroner's lab.

Gloves and gowns or lab coats are required when handling blood. Face masks or face shields are required only for exercises involving splashes and spattering (the arterial exercises included in the next chapter are examples). Face masks should be available to all who request them, for their sense of well being. Plastic aprons are good as they repel blood rather than absorb, and are thrown away after the exercises, but they leave the arms unprotected.

Shoe protectors are a good idea but care should be taken in walking fast over polished floors such as a gymnasiums. Students should be notified to bring a spare set of clothing, including shoes, to class for lab exercises. Bloodstained clothing should be washed as soon as possible with bleach and not placed with family laundry. Some agencies now provide PPE (personal protective equipment) packs with disposable jumpsuits and gloves in the correct size, as well as head, face, and shoe covers. These save time and may be worn for the entire week if the participant follows rules of behavior around the experiments.

If gowns are used and washed each day they should be dried with sheets of fabric softener. The reason for this is that the sheets provide a water-repellent coating to the material. An added protection is to spray clothing with a water seal to keep blood from soaking into garments. Wet blood transfers and is more apt to reach the bloodstream if held in contact with skin breaks.

Having containers of disinfectant, diluted bleach, and spray bottles of diluted bleach available near exercises is a good idea. With bloodstain pattern exercises accidents do happen and will be investigated for pattern characteristics by the entire class, including the instructors. Spraying a perimeter of diluted bleach emphasizes the need to keep a safe distance as well as to begin clean up when curiosity wanes.

CLEAN-UP TECHNIQUES

There are three stages to clean up for bloodstain pattern workshops, and no single treatment covers it all.

- Blood removal
- Occult blood removal
- Disinfectant

Blood removal isn't as easy as it might seem. The reason is that blood is both a liquid and a volume of particles. Any cleaning cloth will pick up the liquid and some, but not all, of the particles. When the cloth is saturated, the effect will be simply shifting a group of red cells around. For this stage many additional absorbent pads are required. Stacks of paper towels, clean terry cloth towels, or very absorbent sponges may be used. A stack of paper towels should be used like a sponge. Start with dry towels, press down over a pool of blood and allow to aspirate liquid, then fold in to scoop up blood cells. Avoid spreading or wiping (it makes a bigger mess than helps). Fold the paper towel to a clean side and again blot and scoop. Discard and start with a fresh pad of paper towels. If a sponge is used, start with a moistened (not dripping wet) sponge rather than a dry one. This hemolyzes the blood cells and aids in removal.

Occult blood removal follows removal of most of the visible blood. Water and 10 percent household bleach are spread over the area and allowed to stay wet for at least 10 minutes. Two applications are good but not required. Bleach should not be used undiluted for two reasons. The fumes of undiluted bleach are hazardous. Even diluted bleach should have good ventilation. Secondly, the antimicrobial power of bleach is less effective undiluted because it tends to burn the outer layers and not reach the layers of debris carrying the microbes. Bleach diluted with water helps both to dissolve debris and assist in killing. Clean up is especially difficult around firing ranges where lead cannot be washed into ground water supplies. The area can be sprayed with aerosol disinfectants, which should stay wet for at least 20 minutes.

Disinfectant is the final and a separate stage in cleaning up a bloodstained area. The reason clean up occurs first is because mold, bacteria, and viruses can be protected within lumps of dried blood. Disinfectant spread over the lumps will affect the surface but not the interior of dried material. Because blood cells are microscopic, just cleaning up the obvious material may still not uncover all hidden hazardous microbes. After thorough cleaning, a disinfectant is applied. Although 10 percent bleach is effective against many viruses and bacteria, it may be less so against others. Several aqueous (dissolves in water) industrial disinfectants are on the market. A thick enough layer of this is added to the area; it should be allowed at least 20 minutes to dry, and an hour is better.

During the workshop it is essential that tubs of disinfectant or 10 percent bleach be available to completely submerge bloodstained equipment. Containers such as cutaway bleach jugs or emptied cat litter buckets should be provided for broken glass and other sharp-edged material to be discarded in separate containers. Magical supply stores have retracting knives, which are good for pantomiming knifing assaults without a sharp edged instrument. Knife blades are common in crime scene evidence and should not be omitted from exercises, but care is essential to control their involvement.

The final needs of a workshop are disposal of cardboard and butcher paper exhibits. All paper products not claimed should be burned. All paper products to be saved should be coated with at least three thin layers of clear lacquer or acrylic spray paint. If the exhibits are to be used in future workshops, it is a

Figure 31-2

Plastic box encased exhibits.

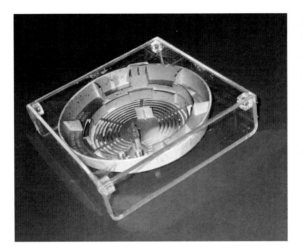

good idea to laminate them with plastic. Page-sized 1.3 mil sheets can be obtained from many office suppliers, and thick 3 plus mil sheets may be ordered from plastic manufacturing businesses and some hardware stores. Encasing in plastic permits passing exhibits around for examination (Figures 31-2 and 31-3) without a need for gloves.

Figure 31-3

Office suppliers and photocopy places have sheets of plastic to heat-seal individual cardboard targets.

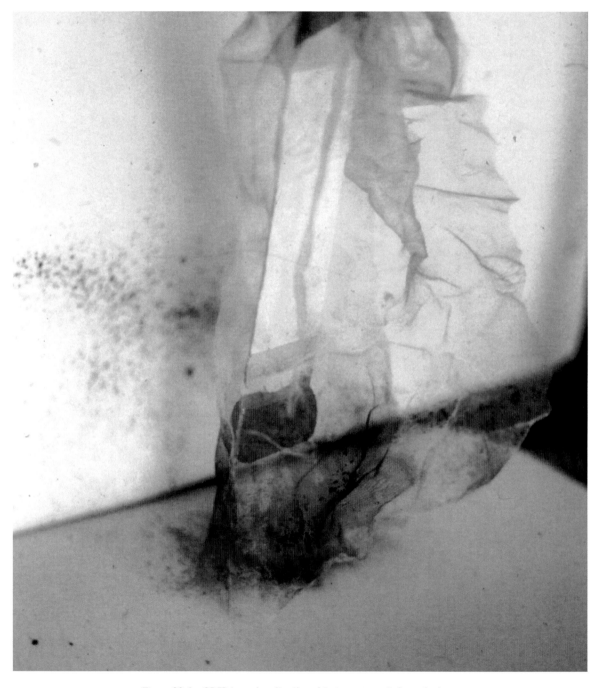

Figure 32-1   GDIS (gunshot distributed impact spatter) through sheers.

CHAPTER 32

SPATTER GROUP EXERCISES

EXERCISES IN IMPACT SPATTERS

Experienced analysts in bloodstain pattern evidence (BPE) know that spatters are only one part of the physical evidence available at the scenes of violent crimes. Still, many workshop participants expect to memorize patterns that identify characteristic events such as gunshot distributed impact spatters (high velocity impact spatter, high velocity spatter, blow back spatter, back spatter), shown in Figure 32-1, and blunt force impact spatter (BFIS, bludgeoning, medium velocity impact spatter). Participants may wait an entire lecture or workshop week for the instructor to cover these subjects while ignoring other potentially valuable material that may also show the considerable overlap in characteristics of spatter patterns. It is not uncommon for students to attend 40-hour workshops just to learn what language to use with respect to "spatters." The first consideration in training thus becomes undoing misunderstandings and subtly but firmly redirecting interest into the whole discipline of BPE. For impact spatters specifically, this involves illustrating that blood spots can be classified as many things, with the same size and shape individual spots present at widely different events.

In forensic science laboratory training, participants learn that evidence may be of two classifications: class characteristic[1] and individualizing characteristics (also referred to as individualizing features). The former, class characteristics, defines evidence as part of a group with the same characteristics. The class can be sequentially limited but will always contain more than one item. The usual example is of a motor vehicle. There are millions within the class. SUV is specific vehicle style but still one of many. A black SUV of a specific automotive manufacturer is further limited, but may be one of many within an region. A California license plate number BF1 2345 is specific for one vehicle upon which the plate is located. Being mobile, a plate can be switched to another vehicle different than the one we want to identify. Noting a black SUV with a broken headlight and blue paint scratch transfers to the right front bumper, having the license plate of BF1 2345 between noon and 2:00 P.M. on Monday 11 December

---

[1]Saferstein, Richard. (1990). *Criminalistics, 4e.* Prentice Hall, New Jersey, 50.

2006 narrows the vehicle down to a specific individual. This latter is a definition in individualizing features, or those characteristics that set one item apart from all others.

A bloodspatter pattern is a class characteristic. Any number of events may distribute a group of blood drops that will be recorded on surfaces positioned along the flight path of the drop array. Just being a group of spots does not immediately identify what event caused initial bloodshed or projected subsequent drop scatter. Class characteristics cannot be used solely by themselves to draw specific conclusions regarding the identity or involvement of a suspect. Limiting characteristics downsizes our class but still does not individualize a pattern.

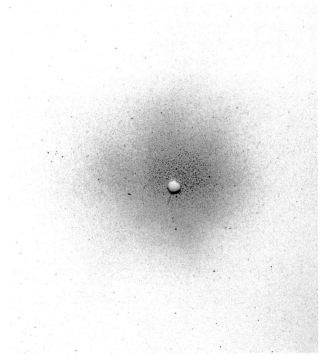

Figure 32-2

Example of mist-sized pattern.

It is important to expose subjective thinking early in training. The power of suggestion can be very strong, so that one member of a team can influence the entire thinking during the week. A way of dealing with this is to start out with an exhibit such as that shown in Figure 32-2. The class is asked how they would identify the pattern shown. Someone will often provide the identification label high velocity impact spatter (HVIS). Then the class is asked, "how do you know that?" Some may respond because of the size of the spatters, and other more cautious individuals will claim a bullet hole (i.e., bullet hole means gunshot and gunshot means HVIS).

This is especially important since incomplete prior training may orient participants toward reliance on information other than the bloodstains to identify spatters from impact. The usual approach is to find out if a gunshot was involved and then label any array of spots as HVIS, or if no gun is involved, as MVIS. Law enforcement officers with homicide experience are not as caught up in the subjective as those inexperienced, but the lesson benefits all.

The class is then shown an exhibit illustrating how the first exhibit was made with an artist spray-paint assembly and a pencil hole (Figure 32-3). When cries of fowl resound it can be pointed out that such theatrics as showing this figure could destroy their analysis in court. The objective of the demonstration is to impress upon the class that they should learn to identify pattern characteristics without trying to memorize what they look like. More difficult course participants must not rely on overheard information, and possibly falling back on subjective identification such as the bullet hole means it is HVIS. The small size helps identify an event that atomizes blood. Gunshot is not the only act that will do that. A case in Section II involved atomizing of blood and CSF without any firearm involved (Figure 32-4; color of Figure 21-8). Also patterns at crime scenes are never as pure as the exhibit. Mist will have fine and small stains within the pattern with an occasional (or even a few) medium and a possible large stain. The shape of the pattern suggests an event directed at the target and limited to a round whole arrangement. This severely limits the type of act that could have resulted in this pattern. The next step is to think in the three-dimensional context in which the pattern could have resulted and work from there. Be extremely cautious with patterns made up of spatters showing no directions of travel and having the group of spots arranged in round and oval shapes without *absence* areas.

Another way to deal with preconceived ideas in a workshop is to provide a scenario target, an exhibit that has been specifically manufactured according to a theoretical crime event.[2] A question is asked of the participants in which the answer can be shown to be speculation, not based on the bloodstains themselves, something like Figure 32-5. Beware of providing cases in any form with law enforcement classes, because participants may focus on solving the case rather than observing exercises. The exhibit should be put away when exercises are in progress.

A laboratory approach to emphasizing variation in spatter size, shape, and degree of impact is to include like exercises within the

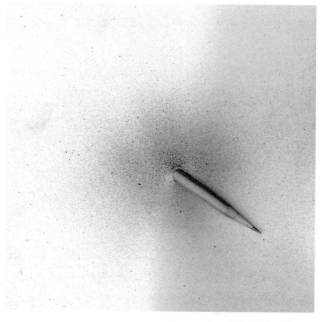

Figure 32-3

Mist pattern construction exhibit.

Figure 32-4

Mist-sized pattern within a CSF/blood pattern distributed under high pressure.

[2]Wonder, Anita Y. (2003). *Journal of the Forensic Science Society Science & Justice*, Vol. 43: 3 July–Sept., 166–168.

Figure 32-5

Floor pattern puzzle exhibit (based on patterns found at the scene of a double homicide).

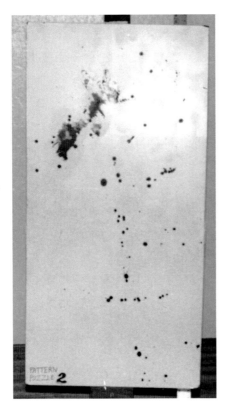

Figure 32-6

Two different spring tensions provide reproducible impact spatters at different levels of force. The boat was light impact (mouse trap spring) and bread board (rat trap spring).

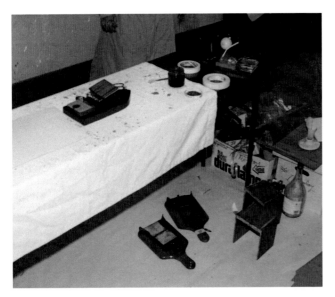

same lab period that illustrate the class of spatters (bloodstains, spots). In the bloodstain dynamics workshops conducted in California, Ohio, and New York, the first day laboratory exercises included two impact situations: a spring trap device impact (Figure 32-6) and a reproducible splash from free falling objects. The spring trap exercise included three different inertial force traps, one with little force, one with strong force, and one with force equivalent to a bullet impact (nicknamed Slow Bullet) (Figure 32-7). Patterns were created with different quantities of blood, variations of dampening effects with hair, sponge, cloth, and positioned at different angles to the recording target, flat, elevated, angled. The objective was to show participants that the same repeated event could provide a great deal of variation between resulting pattern appearances and distribution of blood drops. Teams were encouraged to view the exercises of other teams throughout the week. For this a large gym area worked well.

The splash exercises involved two different free fall devices that could accommodate an addition of weights from 5 ounces to 30 pounds (Figures 32-8 and 32-9). The bottom of the devices had bolts to attach shoes from a child's to a large man's size. The splashes could also be viewed on surfaces at two different angles, flat on the floor and elevated a few inches. A variation included with the exercise provided to make a shoe print with the shoe after it fell into blood. Shifting the weights between two sides on the device helped illustrate weight distribution in steps into blood.

Introduction to blood behavior is also included in exercises examining drop size, distance fallen, and characteristics of surface material in final stain appearance. An addition was made in which light, normal, and heavy blood samples were added to the same exercises. The three bloods were made up from a sample of EDTA freshly drawn blood and adjusted to hematocrits (packed red cells/plasma ratios) of less than 15 percent, 45 percent, and greater than 60 percent. These demonstrated the effects of blood source and bleeder in the appearance of drip cast offs.

Unlike traditional workshops, however, the Bloodstain Dynamics workshop did not use blood dripped onto angled material to train measuring of impact spatters. Large drops of blood involve considerably more overflow, extended edge characteristics, and measurement distortions than true impact spatters. Following the distance fallen, angled effects, and surface texture exercise, workshop participants were given a series of blood-spatters from actual impact spatter targets that were enlarged eight times. The cut-off points for measuring the stains were demonstrated in class. After the participants were familiar with cut-off points, a series of 25 doubled in size bloodspatters from various impact spatter targets were given as homework (see Appendix D). The objective was to train how to measure true impact spatters, and more important show how to calculate angles, with stains that were easy to see but accurate representations of what they would measure at crime scenes. The exhibits were then downsized until the teams were working with actual impact spatters of the size usually encountered. An added benefit was that they didn't need to wait until the drip targets dried before beginning to measure, and they didn't need to take blood-stained material out of the gym to work on it at night.

After the 25 stains were completed, teams were provided time with a

Figure 32-7

Automobile spring and fiberglass support provides impact force comparable to a less than maximum force bullet. Nicknamed Slow Bullet. This was convenient when firearms were not permitted.

Figure 32-8

Free-fall device to which shoes could be attached and weights added to the top pans.

Figure 32-9

Detail for large free-fall device.

photograph of a wall, reproduced to size, where bloodstains appeared to be recorded as BFIS (Figure 32-10). This was listed as a String Reconstruction practice session. There is more discussion of this later with the practical exercises.

A variation that was available sometimes but not used as a specific exercise was a concoction known as Hot Lips. A pump-type bottle was filled to just below the aspirator tube. A few tablespoons of stirred egg white were added to the bottle and the contents gently shaken. The wrist part of a latex glove was attached to the dispensing end of the pump to simulate a mouth. This could be directed as a reproducible cough.

The objective of the devices and different amounts of force was to consistently illustrate that considerable variation exists for impact spatter events. It would be impossible to provide visual examples of every condition that will affect patterns seen at actual crimes. At least in providing an array of variations within the workshop, participants could see that memorizing a specific pattern would not assist in identifying the complex impact events of crimes seen later. Learning the conditions of variance and how each affects the resultant pattern would be much more productive and job related than reproducing the same patterns in each workshop. In California it was common for agencies to send one to three participants to each program. Discussions regarding the actions of previous participants with later ones established that teamwork was occurring with identification, detectives, and crime lab staff conferring to resolve cases. They shared their experiences from the various workshops.

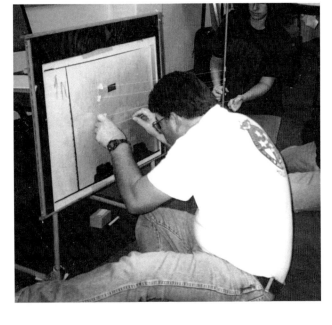

Figure 32-10

Practice target based on an actual to-scale scene photograph.

EXPERIMENTS IN CAST OFF EVIDENCE

It is important for workshop participants to understand that each pattern category has many permutations. The classes of cast offs are characterized by a group of stains resulting from blood drops sloughing or being cast off a bloodied object as it moves. There are basically three ways blood drops may be cast off; thus the variations are named:

- Drip cast off (LVIS, passive stains, gravitation, drip trail), slow movement like walking or running
- Swing cast offs (cast offs, parent and secondary drops), increased velocity over the range of swinging blows or swinging arms
- Cessation cast offs, showing the effect of motion followed with abrupt stopping

The importance of the entire cast off group is that they usually are found in and among frequently overlapping impact spatter patterns. The individual stains that make up the patterns need to be recognized as originating from separate events. Impact spatters lead back to a common origin in time and space, and cast offs define an object's path over a range in time and space, along which blood drops were distributed. Because cast offs do not originate from a common origin they are not measured and never should be included in a reconstruction of the origin of an impact. They may be used, however, to locate a general area or relative positions where the assailant stood or moved during the swings of a weapon.

Drip cast offs are the simplest form. Blood on an object is thick enough and still liquid to an extent that it can flow and separate into drops. The carrier of the blood source may be stationary or moving at walking speed up to a fast run. The exercise that commonly is used for this pattern type is to have a partici- pant carry a dripping blood source along a strip of butcher paper or cardboard sheets placed on the floor. The initial blood source was usually a dropper into which blood had been siphoned prior to the action. Squeezing on the drop-

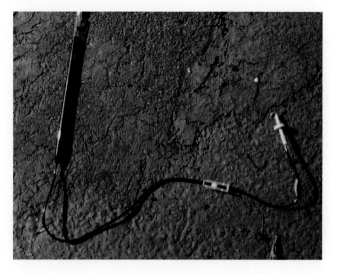

Figure 32-11

The two-way venting system for transfusion pouches is an improve- ment over the hand-held droppers.

per could be variable and jostling emptied the blood prematurely. An improvement was to use blood bank units, pouches, with an IV two-couple line attached (Figure 32-11). The rate of drip could be adjusted and the partici- pant only needed to hold the drip line while walking or running. There was the added ben- efit that blood did not need to be handled to transfer to containers for other exercises.

To show permutations from simple drip cast offs (passive stains, drip trails, gravitation stains), each of six teams was given one vari- ation to perform. All teams were responsible for observing the results. This also helped in having students observe patterns that they did not create themselves but were to describe later. The variations demonstrated:

1. Effects on a drip cast off pattern by blood coming off a carried heavy body. In some classes a lightweight volunteer was carried while holding the drip line; in others a dummy was carried, and in a few classes a 50-pound-size bag of cat litter was carried. The objective was to compare an easy walk/run with a stressed struggling with a weight movement.

2. A variation of rate of bleed was included to show the amount of blood dripping affected the overall pattern appearance.

3. Two members of a team were instructed to struggle and act out a fight over the path of the dripping. This showed participants a struggle composite where shoe transfers, drips, and swing (or flicking) cast offs occurred along a line of the drips.

4. Participants were instructed to dip objects in blood and carry them over the length of the recording surface. These included a wig, knife, wood board, newspaper, Teflon® rolling pin, and terry cloth towel. This showed how drop size and width of the pattern might vary per object being carried as well as the fact that blood drops decreased along the length of the target, unlike the patterns where blood was continually added from the drip line (simulated injury).

5. The participant fakes a drunken stagger while passing over the recording surface.

6. The drip line is held in one place to create a pattern known as blood into blood.

For swing cast offs the traditional training technique is to provide a room or scaffold enclosing a space roughly 8 feet high and extending 8 feet wide with a depth of at least 4 feet (Figure 32-12). A variety of weapons are provided and participants are allowed to hit a blood-soaked sponge, swinging the weapon back and forth in overhead or horizontal blows. It was noticed early in workshops that leaving this entirely up to the participant could result in a misconception. People in a bloodstain workshop wanted to create bloodstain patterns. Hitting the bloody sponge wasn't important. The swings with the weapon gave emphasis to the back swing not the forward swing. This resulted in a strong snap at the reverse, farthest from the victim. The forward swing toward the blood (victim) was less forceful. The patterns reflected a situation in which the majority of blood drops were cast off on the reverse swing and the snap took off most of the remainder. The return to the blood source (victim) resulted in almost no blood cast off.

Figure 32-12

A scaffold is the best way to study swing cast offs. The space should be at least 8 feet high with sides 6 to 8 feet deep and 10 to 12 feet wide. A revised model for later workshops was made of polypropylene pipe with shower curtains to enclose and contain the space and blood. A roller could be fitted with standard butcher paper rolls that were changed between teams.

In assault cases the action is reversed (Figure 32-13). The force is in the forward blow (toward the victim) and the back swing less forceful because it is less important than the assaulting action. The back swing also has more blood on the carrier, thus bigger drops are available to be cast off. To help duplicate the types of patterns participants will actually see at crime scenes, the classes were prepared by instructing them to ignore the creation of bloodstain patterns. They were to visualize someone or something for which they felt a strong emotional dislike. The visual images most used were ex-wives and commanding sergeants. Ex-husbands and higher ranking officers were less effective in creating the right mind set for the exercise. When the participant used the weapon in an emotional assaulting manner, the strongest pattern was toward, not away from the victim. With weapons that did not hold blood well, nothing might come off on the back swing, but a few drops come off on the more forceful forward swing. A crucial observation is that cast offs are characterized by direction of travel toward the victim, and impact spatters have directions of travel always away from the victim.

The paddle fan device known to many workshop graduates was deleted from the author's program in 1987 because of the confusion it seemed to create in participants. Essentially this device is a high-speed cast off example, yet often was used to illustrate impact velocity. Some benefit is derived by pointing out the distance drops traveled from the motor blades and the size distribution of the measured stains. Usually an array of sizes could be found close to the mouth of the device with a decrease in the number of different sizes found as one progresses away from the fan. Few instructors provided magnifying glasses to the class, thus the smallest stains were not always appreciated. The main point was that a calculated speed for the fan was a relative speed for cast off drops from the quantity of blood deposited onto the fan blades. This information could not be used to conclude a similar speed for an impact, such as gunshot, nor the speed of individual drops. Unfortunately students often saw the experiment that way.

In place of the paddle fan, an experiment with cessation cast offs was used. A block with a raised tubular steel bar was used to demonstrate defensive gestures and cast offs resulting from immediate cessation of weapon (bloodied object) motion (Figure 32-14). The student would hit a bloodied piece of carpet or sponge, then hit the bar without touching blood to it. The weapon would stop but blood

Figure 32-13
An exam question asked where the victim was located, at position A, B, C, or D (letters not shown). The larger drops are away from the victim because that's when the weapon has the most blood. The smaller spatters (spots) with more marked directions of travel are toward the victim because that is where the force of the blow is intended. Workshop participants sometimes reverse the emphasis with a snap on the reverse so that little or no blood comes off toward the victim.

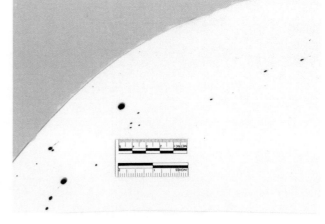

Figure 32-14
An exercise with cessation cast offs can follow the swing action by striking the bar at a position below where the weapon has become bloodied. This can demonstrate self-defense gestures that are very common at the scenes of beatings.

Arterial damage patterns: Bloodstains resulting from the distribution of blood from a breached pressurized blood vessel.

drops would be distributed beyond. These might show a relative shape and size of the bloodied object (i.e., investigative leads). This showed how one could use defensive gestures of a victim to identify the weapon used in a beating.

A technique not used in our workshops but one with potential to increase the information participants would have is to provide a vertical bar supported by a heavy base and instruct participants to swing batter's style at the bar. This would give good examples of batter's swing cast offs as well as cessation cast offs when the weapon stopped at the victim (i.e., in defense gestures). The reason it was not done in our workshop was the need for protected walls around the batter. Most of the author's workshops were held in gyms without separated experiments and sufficient enclosures. The scaffolding used for cast offs did not provide sufficient protection.

EXERCISES IN ARTERIAL DAMAGE (PROJECTED BLOOD) PATTERNS

Most 40-hour workshops do not include specific **arterial damage** (projected blood) **pattern** exercises. The patterns are regarded as included in splash with volumes of blood dropped or thrown onto recording surfaces placed on the floor. This may lead to a misconception that arterial damage patterns are always large quantities of blood. Some arterial projections can involve no more blood than a single column of drops leaving a row of spatters that resemble a swing with a bloodied weapon (i.e., swing or cessation cast offs). The omission of specific exercises during workshops is understandable due to the amount of blood necessary to prime a device for demonstration, and the level of control needed to contain biological hazards. Still, the importance of these patterns and the frequency of encounters with the evidence during casework suggests a definite need to see the results of arterial dynamics first hand.

Five subcategories have been defined:

1. Arterial Breach.   Identifies where an artery is initially breached. As pointed out in *Blood Dynamics*,[3] breach may include severing of the artery but often does not. A

---

[3]Wonder, *Blood Dynamics*, 139.

hole or puncture occurs via an aneurysm with blood pressure projecting a column that separates into blood drops after leaving the vessel. Because of arterial physiology, the hole may close if constriction follows rapidly, or be plugged with bone or clot material. This latter prolongs life and stops loss of quantities of blood. Patterns locating a breach may be indistinguishable from an impact, or swing or cessation cast offs. Two things are necessary to position an arterial breach: knowledge that an artery was injured and presence of patterns characteristic of arterial spurt, gush, rain, or fountain at the location or leading away from it. If the breach is small and the blood pressure high at the moment of breach, drops separating may be responsible for mist-sized spatter.

2. Arterial Spurt.   A term commonly applied to a series of separated spots in a linear or narrow rectangular arrangement showing some undulation to the whole array. A pulsing artery may project several of these over the course of bleeding. If the injury is to a minor artery, constriction and an eventual clot plug may stop bleeding. Statements that arterial injury assures death are not necessarily true although arterial injuries tend to have more serious consequences than venous injury. Conclusions of life and death must be made through a qualified medical practitioner. A special note is that arterial spurts were used in a medical science technique prior to 1911 (Figure 32-15). Their pattern recognition has science validations for arterial damage bloodstains dating from before 1911.

3. Arterial Gush.   A term used interchangeably with arterial spurt, but viewed more as unseparated spots. This would occur when the projection of an arterial stream does not separate into drops. It is with arterial gush that a phenomenon between Newtonian and non-Newtonian fluid behavior may find applications in BPE. This was pointed out in *Blood Dynamics*,[4] illustrating gush patterns of each fluid type.

Blood flow is normally non-Newtonian throughout the body. Flow around the brain and within the heart, however, is Newtonian. When the body is threatened from trauma, extreme fear, stress, and/or exertion, 80 percent of the blood flow is directed to the head and the heart[5] with Newtonian flow between. Cutting the carotid artery during this time may result in a projection of a gush with satellite spatters around the core column. The pattern has been described in casework and confirmed as the blood type of a victim who suffered carotid arterial severance. Laboratory reproduction has not been reliable as of yet but study continues.

Figure 32-15

Arterial spurting bloodstain patterns were known to medical science prior to 1911. This is a sketch of a pattern resulting from a dog tibial artery spurting at a rotating paper drum.

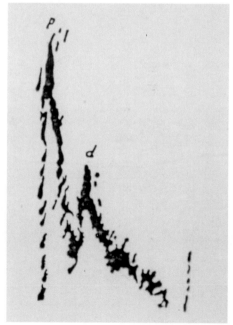

---

[4]*Ibid, 74.*
[5]Sohmer, Paul R., MD. (1979). *The pathophysiology of hemorrhagic shock.* In *Hemotherapy in Trauma & Surgery.* AABB, Washington D.C., 2.

Arterial fountain: A vertical pattern formed from the fallout of an artery directed upward.

4. **Arterial Fountain.** Patterns that result when a projected column of blood is aimed up. Drop separation results in a cascade down by gravitational force. If the action occurs near a wall or vertical recording surface, a fountain- or bell-shaped image may be seen. This can be confused with a fountain distribution by flicking a bloodied object at a surface. The difference is seen in the uniformity of the stain sizes aligned in parallel arrangement with arterial and in a range of stain sizes all at slightly different angles to each other in cast offs.[6]

5. Arterial Rain. An identification that can be recognized alone in the absence of other arterial damage patterns. Breach and fountain usually are recognized after arterial spurt or gush is identified. Rain occurs from fallout onto horizontal surfaces from a pulsing artery and is readily identified as randomly scattered small to large round individual stains. The pattern may be linear if the fallout is from a single projected spurt or gush, or scattered over an area if the fallout is from a fountain or the victim moves around during opened arterial pulsing. The importance of rain, as with spurt and gush patterns, is that it can indicate the existence of arterial damage when the autopsy fails to mention it. Interview or cross-examination would be indicated to establish that arterial involvement occurred in the absence of mention in the autopsy report.

Three exercises have been used to illustrate arterial damage patterns for BPE workshops:

- Recirculating pump device
- Video of the pump device used in an enclosed area
- Syringe projection

Figure 32-16

Recirculating fountain pump with one-way float valve, used in arterial damage exercises.

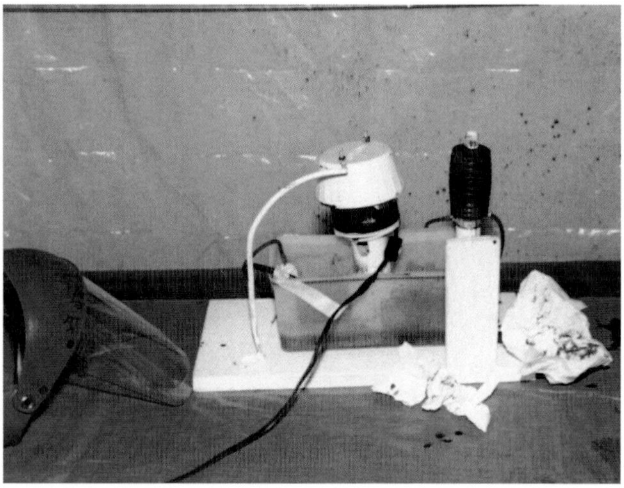

A recirculating pump device was developed after attempts with various fountain designs. The pump was a small fountain used for water features in gardens (Figure 32-16). Latex tubing contained the flow circulating between the pump and a reservoir. A minimum of one liter, two full whole blood pouches, was required to prime the pump. Packed red cells undiluted would clog the system and interfere with flow. A dilution of 2 parts red cells to no more than 1 part physiological saline (or contact lens saline) was acceptable for the reservoir.

This system could not provide the positive pressure necessary to project an arterial stream characteristic of a victim being assaulted. To achieve this a one-way flow

---

[6]Wonder, *Blood Dynamics,* 76.

valve with a hand siphon bulb was inserted in the tubing line. Workshop participants were divided into four stations: the motor operator, the heart person (siphon bulb), the victim (tubing holder who projected stream toward the recording surface), and observer/notetaker. The system was extremely messy, very much like an actual scene of arterial damage, but impractical to continue (Figure 32-17).

Figure 32-17

The exercise required groups of four participants to keep blood under control and was always messy.

British instructors followed this technique and improved upon the model (Figure 32-18), but used a different design, a peristaltic pump with an oscillating arm, which was controllable in frequency and stain size. They included a manifold device with a number of channels where tubing could have different sized holes. Pressure and size of breach could be compared. It could also be handled by one or two people (Figure 32-19).

Devices in the usual chemistry laboratory include simple peristaltic motors that force fluid through the line by clamping off the flow behind. The effect resembles arterial pulsing but is a shutting off rather than a periodic increase in pressure (Figure 32-20). Still, the patterns are similar to actual arterial events and the difference is considered insignificant compared to the increased control and composition of the device. Thicker fluids (i.e., blood products with a higher ratio of red cells) may be used in a simple peristaltic pump. The importance of illustrating arterial type patterns can be seen in the volume of casework where arteries confuse the analysis.

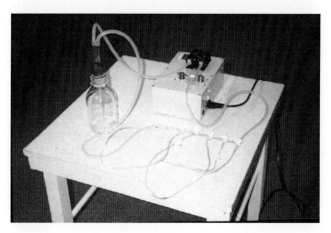

Figure 32-18

Adrian Emes and Chris Price improved upon our model and used it in many workshops in England, Europe, and Asia.

A video of the pump device became necessary when the workshop venue included only open gym facilities and whole blood pouches were unavailable. Prior to the class, a location was found where the pump device could be set up and the blood projection contained and cleaned up thoroughly. These areas can be a shower stall (Figure 32-21), an unused toilet stall, or an enclosed area in a water-safe location. Butcher paper and large sheet white cardboard was placed around the area of the pump. The different positions and actions possible with a theoretical victim moving with a punctured artery were videotaped.

Figure 32-19

The modification helps with the mess and doesn't subtract from the fun.

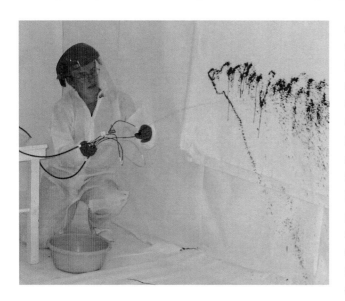

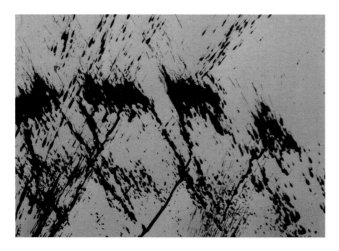

Figure 32-20

Common peristaltic pumps show undulations but from the pressure being stopped in between cycles not increased as in a true arterial function.

The lab session was then conducted with the video being shown, followed by the participants examining the paper and cardboard records of the videotaped exercise. This provided several advantages over the actual use of the pump although not as entertaining as projecting blood.

Syringe projection was used only in specific situations, a part of the workshops added at the request of participants as an experiment in reconstruction. This began as an extension of the transfer pattern exercises, but became a combination conclusion of the workshop as a whole. Students were given free reign to choose items to use in a scenario of their choice as long as the experiment included different overlapping events occurring after time lapses. Some teams requested arterial damage pattern projection. By the time this experiment was to be done, the device and blood spill had been cleaned up, and the video and targets put away. To provide the teams with evidence for their scenario a blood-filled syringe, usually a 30 or 50 cc with a 10 gauge bone biopsy needle, was used to project the required stream (Figure 32-22). This was done by an instructor not a participant. The effectiveness of the stream to separate into drops did not necessarily require a needle attached to the hub lock. The patterns can resemble arterial damage stains as the pattern on the seatback in Figure 32-23 shows.

Attempts to find actual heart pumps to use in very realistic arterial projection were not successful, due to the competition between manufacturers of these devices. Heart surgery and supplies of implanted heart pumps are a lucrative business with companies adamant to keep their own designs secret. Attempts to acquire a device or to even learn of the essential components was met with suspicion and communication blocks. In the future government agencies may be more successful in finding a realistic design for reconstruction casework.

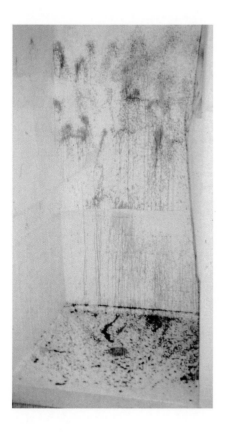

Figure 32-23

A simulated arterial damage pattern in a mock crime scene, vehicle.

Figure 32-21

When clean up and blood availability became an issue a video was substituted. The demonstration for the video can be made in a shower stall for easier clean up.

Figure 32-22
20 or 30 cc syringe used with broken off needle works well, or remove needle from the luer lock hub.

Figure 33-1   Various targets with the objects used to investigate simple direct transfers are laid out on the gym floor.

CHAPTER 33

EXERCISES WITH GROUPS OTHER THAN SPATTERS (NONSPATTER GROUPS)

EXERCISES IN TRANSFER PATTERNS

In contrast to spatter groups, which are class characteristics, transfer (contact, compression) patterns have the potential to have individualizing features. This means that it may be possible to connect patterns in blood to a specific event, material, assailant, point in time, or sequence of actions. Since trace evidence usually is handled by crime labs, law enforcement investigators may put less emphasis on training. This is understandable, but not advisable since the transfer category of investigative transfer (contamination), from early investigators accessing the scene, can invalidate some important evidence for court presentation later. Sharing with participants how blood can be unique in recording not only the individual items present, but how they moved, were used, and interacted during and after the crime. This can train investigators to note, protect, and seal areas of the scene for extended processing later, and in the meantime provide potential investigative information useful during interviews.

As with other pattern types, transfers can be subdivided into subcategories:

1. Blockage transfers (voids) are patterns created when one object (obstruction) acts as a **template** or **stencil** to block the distribution of blood within a spatial area. If the obstruction is moved or removed, the blockage outline may be sufficient to identify what was present during the distribution of blood. Even if the blockage transfer isn't of sufficient detail to identify the object, the nature of the transfer may help locate direction to and level of the blood distributing event that outlined it.[1]

    These should not be confused with absence, which is the lack of bloodstains because of angles and locations between the blood source and the recording surface, not involving a blockage, i.e., the blood would not be deposited in an absence area whether or not a blocking surface was present. An example of blockage would be if a drinking glass was sitting on a table near a victim of gunshot. The glass would block blood spatters and might

Template (stencil): An object that acts as an outline or blockage when spatters are recorded.

---

[1] Wonder, *Blood Dynamics*, 85.

be identifiable from the ring left on the table. Spatters projected beyond the glass to each side would have areas of no spatters too, but they would be the result of the cone of projection from the shot, not an obstruction blockage (i.e., an absence pattern).

An interesting effect was seen at a multiple homicide by knifing (Figure 33-2). Is this a blockage or an absence? The way to label the pattern where there are no or noticeably fewer spatters, is to first ask if you removed something, would it then change the amount of spatters in that area. If so, then the classification is blockage, and is interpreted that way (the front edge is blocking the reverse side, possibly). If removing something wouldn't make any difference then it's an absence, and dependent on conditions other than an obstruction (i.e., manner in which the shot was fired or beating delivered, etc.). So, the spatters on the adjacent surface to the left would have the same amount of spatters if the frame were removed, which means the *void* is a positional one and is labeled as an absence in that respect. The significance is that an absence can be used to determine angles to the source of the spatters, while a blockage identifies changes to the scene.

Because blockage transfers require a primary event to distribute blood in order to form the blockage, training is conveniently coupled with spatter experiments. This may be with either the spring trap device or the splash free-fall device (Figure 33-3). An object (bottle, cup, book end, fake gun, knife, belt, figurine, picture frame, empty beer can, liquor bottle, etc.) is placed between the recording surface (cardboard target) and the blood distribution system (trap or free-fall device). The advantage in coupling exercises is to show spatter type bloodstain patterns in a more practical yet complex context as well as emphasize the possibility of overlapping patterns.

Figure 33-2

Although the side without any blood could be considered a blocked pattern, it is also an absence due to the angle it was positioned at the moment of the knifing. As such it can be used to triangulate to the origin. At least one contact with the knife occurred while the victim was farther to the right facing the door frame.

2. Simple direct transfers (compression, contact) evidence requires only brief contact between a bloodied surface and a blood-free surface. The action may be blotting or transfer depending upon the materials used. In a workshop the usual training approach is to provide objects (toys, fake guns, books ends, figurines, knives, various possible blunt force weapons, items of clothing) and fabric swatches (burlap, rug section, terry cloth, raised surface wall paper, muslin sheet, denim jeans, reversed side suede to mimic skin texture, natural hair wigs). Participants are given an opportunity for creativity in dropping, blotting, rolling objects, and touching fabric to the object, then blotting on a recording surface. This may include attempts to reproduce patterns from the participant's cases.

It is important that participants are instructed to use various types of contact, not just simple direct transfers. For example dropping, rolling, and wiping all are direct contacts but provide different interpretation and actions even with simple contact. See if you recognize the common items in Figures 33-4 through 33-9. The answers are at the end of the chapter.

3. Moving transfer evidence is unique for blood because the composition includes a clear colorless liquid and a red pigmented ratio of particles. The liquid may be aspirated but the red cells will aspirate only with sufficient liquid present. Anyone who has cleaned up after a workshop or experiments using blood knows the behavior of blood. Instead of the red cells being removed, they are

Figure 33-3

Showing the importance of blockage with impact exercises trains two things, but also assists encouraging participants to broaden their focus when examining an impact spatter pattern.

often shifted around. Different amounts of pressure when making moving transfer can be recognized. Participants should be told to use the objects available in actions including push, drag, slip with pressure, and slide with changes in direction. Thus, three forms of movement can be further differentiated:

a. Wipe is a pattern where a nonbloodied material passes along a previously bloodied, still moist surface. Red blood cells are pushed along and tend to concentrate on the lift-off edge. The lift-off edge is usually a more even stain with the inclusion of some texture characteristics of the material used in wiping. Wiping over spatters shows shells of the original spatter stains with spreading of the stains in the direction the wiping material moved.

b. Swipe is a pattern that is created when a bloodied material moves against a nonbloodied surface. In this situation blood is removed as the action proceeds, thus the lift-off edge is lighter than the first contact edge. Characteristics often are described as feathered edges.

c. Smudge is sometimes defined as a **moving contact stain** in which the direction of travel and/or nature of the stain cannot be determined. Smudges often are stains that began as either a wipe or a swipe but direction of movement and which surface was bloodied first changes with continued action such as moving back and forth, or in circles over the same area.

Moving contact stain: A transfer pattern that has not been classified into Swipe or Wipe.

A caution is made with training of wipe. Participants may start with wipe but end movement of the cloth beyond the prior deposited spatters (i.e., carrying a wipe until it becomes a swipe). When they attempt to examine the darker leading edge they are actually looking at a swipe, a lighter lift-off edge. Because of the confusion with using and understanding wipe and

Figure 33-4

A.

Figure 33-5

B.

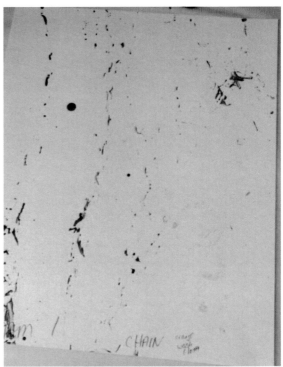

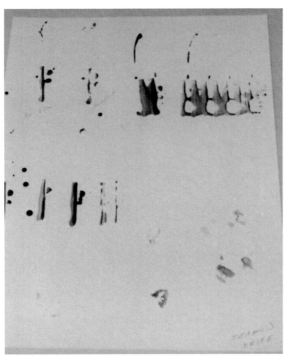

Figure 33-6

C.

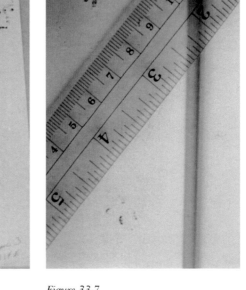

Figure 33-7

D.

Figure 33-8

E.

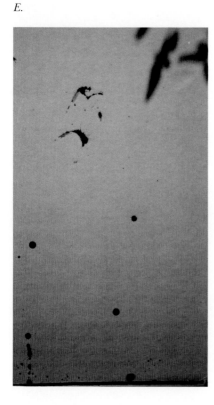

Figure 33-9

F.

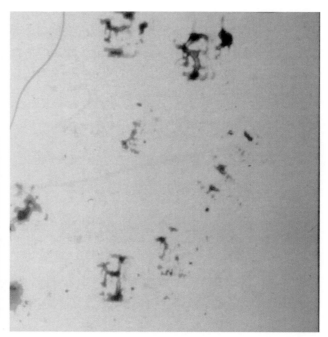

swipe patterns, the term moving transfer was developed for primary training classes.

A special circumstance exists with skin swipes. The principle with swipes and wipes is that blood is deposited or moved along with some absorbed into fabric or other material. With skin, there is no absorption. Skin oils also assist in creating a smooth spread with blood concentrated on the lift-off edge regardless of whether it is a wipe or a swipe.[2]

The combinations of materials, fabrics, and objects seem limitless but instructors should keep in mind some combinations to provide when collecting equipment for the experiments. Provide examples of:

- Something absorbent and something nonabsorbent
- Something with a smooth texture and something with a rough texture
- Something with raised pattern and something with a uniform smooth weave
- Objects with right angled corners and with rounded surfaces

Note defects and areas of wear on materials and objects. Also, point out that the amount of blood has a lot to do with the final transfer patterns. Less blood

[2]Wonder, *Blood Dynamics,* 94–95.

often results in greater recognizable detail. Too much blood may flow together and fail to retain a specific pattern.

Something that was not generally included in the workshops but is a fundamental observation for actual casework is to provide a dress form or wire dummy so that fabrics can be laid across curves of a body. This helps in recognizing patterns later in clothing examination.

EXERCISES IN PHYSIOLOGICALLY ALTERED BLOODSTAINS (PABS)

PABS patterns may be essential to reconstructing the events that left blood at a crime scene. Most workshops include some kind of exercise but not necessarily under the broad heading of physiological change. The importance of illustrating the classification is to inform investigators to specifically look for changes in the bloodstain patterns over time as well as note the stage existing when initially viewed. Changes occurring during an investigation assist reconstructionists to establish the sequence of events involved. The fact that blood is a respiring, physiological fluid that behaves in unique ways helps to identify the suspected substance, which should always be tested to verify it is blood.

Four physiological changes were discussed in *Blood Dynamics*:[3]

- Settling
- Drying
- Clotting
- Mixing with other substances

PABS/settling has been observed at the scene of a double homicide, and since the publication of *Blood Dynamics*, mentioned by readers who have encountered the evidence. To date no case evidence has suggested that settling of red blood cells had value in an investigation. Since most workshops use out-of-date transfusion pouches, settling of the red cells is readily apparent. Transfusion is now almost exclusively with packed red cells where plasma has been removed. Settling is less noticeable. In normal situations blood may not settle in the time it takes to clot or dry.

PABS/drying bloodstain patterns can be seen without any specific experiment. Workshops are increasingly including reconstruction experiments where participants think up their own scenarios and attempt to reconstruct the dynamics for the benefit of the other participants. A requirement for these should

---

[3]Wonder, *Blood Dynamics*, 101.

be that each scenario includes a sequence of time with some overlap between bloodstain patterns. Drying before the next event can be essential in evidence of contamination or crimes over long duration. For economy of time instructors may prefer to use exhibits from the class where bloodstain "accidents" occurred before the entire pattern dried. Another way to include the observation with other exercises is to set up a blood flow and note drying time to decrease and stop further fluid movement. Unfortunately such exercises require considerable time, a specific area for use, and advanced planning.

PABS/clot, shown in Figure 33-10, has been recognized for many years, although possibly not thoroughly understood until recently. As more investi-

gators become aware of the reconstruction potential, more studies are being carried out to understand coagulation at actual crime scenes.[4] Drawing a volunteer's blood and using it in an exercise is the usual approach. If the blood is drawn by venipuncture into an anticoagulant, chalk dust (blackboard marking type) must be added to inactivate calcium chelation and allow some degree of clotting. If the blood is drawn without an anticoagulant, the sample needs to be used immediately in whatever exercise is planned. Instructors should be aware that micro-clotting, which is detected in a clinical lab, is not detectable to workshop participants. Clotting times may increase for actual crime situations because of

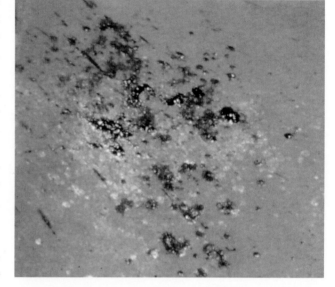

two factors: disruption of the clot during beatings, and absence of tissue trauma (sharp knives and bullet wounds), which prolongs clot initiation.

Figure 33-10

Clot material from a crime scene.

As pointed out in *Blood Dynamics*, a problem with workshop experiments is that procedures require continually disturbing the clot. This disrupts the clotting process and prolongs clotting, with clot recognition difficult at the end. A suggestion for instructors is to do a modification of a Lee White Bleeding time.[5] Instead of using three tubes, use three pools of fresh drawn, no anticoagulant, blood placed on a clean sheet of glass (or better in separate clean Petri dishes). Leave two alone while testing the first pool with a wooden applicator stick. Note the first stage of clumps of red cells on the stick. Switch to a glass stirring rod and test the second pool until it slides aside as an intact lump of cells.

[4]Laber, T.L. and Epstein, B.P. (2001). *Substrate effects on the clotting time of human blood. Canadian Society of Forensic Sciences* 34: 4, 209–214.

[5]Brown, Barbara A. (1993). *Hematology: Principles and Procedures, 6e*. Lea & Febiger, Philadelphia, 215–216.

Leave the third pool until it retracts, approximately one hour. A covered dish (Petri dish with lid) helps prevent drying in the last stage.

The other factor that differs between experiments in a workshop and what happens at actual crime scenes involves the method of drawing blood. Injury to surrounding tissue caused by blunt force trauma or entrance of a bullet involves the release of fluids, which interact with the clotting mechanisms. These may speed up clotting (i.e., shorten time necessary for clot recognition). On the other hand, blood diluted with CSF, tears, and copious blood loss will prolong bleeding and may even completely inhibit clotting.

Some variations to the clotting experiment can be beneficial to workshop participants:

Figure 33-11

Drop of blood that clotted 10 minutes before being dropped on a cotton muslin sheet of low thread count (150).

Variation 1. The surface upon which clotting is observed may be changed to wood, metal, rusted metal, and fabric. (Absorbent materials prolong clotting, but speeds drying, and rust on metal will enhance clotting (decrease time for clotting to initiate) and speed retraction.

Variation 2. The pools can be of different sizes (clotting is independent of volume but drying is not).

Variation 3. Tilt a rough surface just to the point before a pool of blood will flow (when the clot retracts the serum flows downhill because of the loss of cohesion to the retracted red cells).

Figure 33-12

Blood and water mixed and dropped on a sheet of low thread count cotton muslin.

Clots deposited on fabric may sometimes be confused with blood and water mixed on fabric. Compare Figure 33-11 with the next illustration in Figure 33-12.

PABS/mix can be a challenge to the instructors of workshops in how to duplicate mixtures without creating health hazards and clean-up headaches.

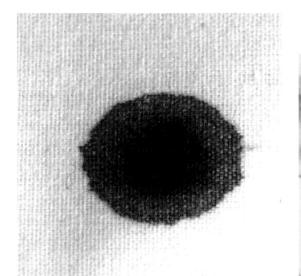

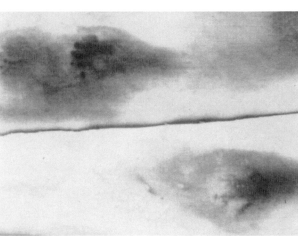

Again the category can be subdivided into specific types of mixes and their interactions:

PABS/mix-water is probably the easiest to illustrate and most common possible mixture at a crime scene. The main factor is to show participants how hemolysis of red cells affects the appearance of a final stain. It should be pointed out that direct view of the stain may not be as apparent as photographs taken and viewed later. The eye and brain compensate for light scatter better than a camera. As shown in *Blood Dynamics*,[6] stains resulting from hemolyzed blood are translucent with the shine of the background material visible in photographs of the dried stains. For demonstration during the class, all stains should be permitted to dry completely before focusing a flashlight beam obliquely on the stains. The shiny appearance for water diluted mix can be contrasted with dusty appearance of physiological watery fluids. Blood and water mixes on fabric may also have a beige band around a dark center with a narrow darker brown irregular edge around the circumference.

Both PABS/clot and PABS/mix-water have been described as target-like stains with a dark outer ring and lighter inside. Clot is often more of a target in that there is another darker area in the center providing three areas—dark, light, and dark. PABS/mix-water by contrast is dark, usually with an irregular edge and inside with a more uniformly lighter, often beige interior. Testing to confirm that the substance is blood, human blood, and that of the victim is always advised.

PABS/mix-physiological watery fluids such as cerebral spinal fluid, tears, joint fluids, and trauma or diseased urine (high protein content) may dilute blood but not hemolyze the red cells. Bloodstains are pink but appear dusted or gritty in photographs of the dried stains. The appearance can be duplicated in workshops using physiological saline (0.85% sodium chloride solution, weight/volume, or contact lens wetting solutions, without preservative).

PABS/mix-miscellaneous includes anything that might feasibly be at a crime scene and mix with blood. One case involved cat litter, and several have involved mucus such as saliva and vagina fluids. Blood is mingled with the substance but not diluted nor hemolyzed. Examples of this may be seen using normal egg white. Gently stir the egg white to break up clumps, and add an equal volume of blood in a screw cap jar. This can be dripped or used with the spring trap device to create   patterns resembling sneeze and cough.

It is essential that workshop participants wait until the various mixtures are completely dry before observing, and that they be warned that photographs

<hr>

[6]Wonder, *Blood Dynamics*, 113–114.

may show better detail than observed with the unaided eye. The camera captures exactly what is there while the human brain may cause focus to miss subtle differences. Having photographs available as exhibits is helpful if participants are disappointed in not being able to see the contrast with their own experiments.

EXERCISES IN VOLUME (POOLING) BLOODSTAIN PATTERNS

Several techniques for this type of training were presented in *Blood Dynamics.*[7] Too often volume evidence is used in a subjective manner. A person may testify that "blood everywhere" is expected. Any opposing counsel that does not challenge this assumption with a qualified expert, if necessary, is remiss in their obligations. In practice many cases involve very little exposed blood. Injuries to the chest and abdomen under clothing, especially several layers, may provide no blood spatters and minimal flow and seepage from the wound. Bleeding in these circumstances is often predominantly internal. Whether there were large quantities of blood distributed or not depends on the injury and may be best confirmed with autopsy information rather than bloodstain patterns. On the other hand cuts and injuries in certain areas such as the face and head may bleed copiously and not be life threatening. An essential part of training in volume (pooling) evidence is instructions limiting conclusions regarding victim survival.

In the author's experience with workshops, current methods to determine the volume of an unknown volume (pool) of blood may be unreliable. Results were considerably varied with some teams giving unrealistic estimates and others getting close to the correct volume. Performance seemed very random, and did not reflect the background or experience of the team members. Still there are cases where no body is found and the only indication that would validate an investigation into a disappearance is that an area suggests too much blood to not be worthy of further investigation. Unlike the other exercises, which provide an exhibit to be recognized later, the volume exercise must be repeated with specific case parameters for each crime investigation. Exercises of this nature are unlikely to be conducted by law enforcement investigators outside training classes. With the need for economy in blood products, this exercise could be deleted.

A favored technique to be used with specific investigations is to add blood to material with the same parameters as the case. It is essential to include all parameters, for example, type of flooring under carpet, type of carpet pad, and type and relative age and wear of carpet. Whole blood is poured onto the carpet

---

[7]Wonder, *Blood Dynamics,* 119f.

system. Whole blood must be specially obtained because transfusion services no longer provide whole blood for normal transfusion purposes. Packed red cells usually provided will not produce the same estimate of volume as whole blood. Blood is added until the same surface area as the case exhibit is covered. This method provides a rough idea of how much blood would cover a specific area of materials. It does not take into consideration the injuries, rate of bleed, and amount of blood available for the bleed before clotting would occur.

Temperature and clotting will affect blood flow and absorption. If an actual case involved extremes in heat or cold this would affect the parameters used in the experiment. The ideal application of this type of training would be if a participant has a case in which they need information. They could supply the parameters for the experiment to benefit the entire class.

ANSWERS TO FIGURES 33-4 TO 33-9

Identification of patterns from simple direct transfer targets:

A. Industrial size pad lock (note one imprint is with the lock inside a man's sock)
B. Plastic encased bicycle chain
C. Knife (predominantly the handle)
D. Mouse tracks on a washing machine (volunteer free range mouse visited the mock crime scene the night before the workshop)
E. Drip and cessation cast offs with a crowbar imprint
F. Officer's badge used in transfers (actually a toy badge was substituted)

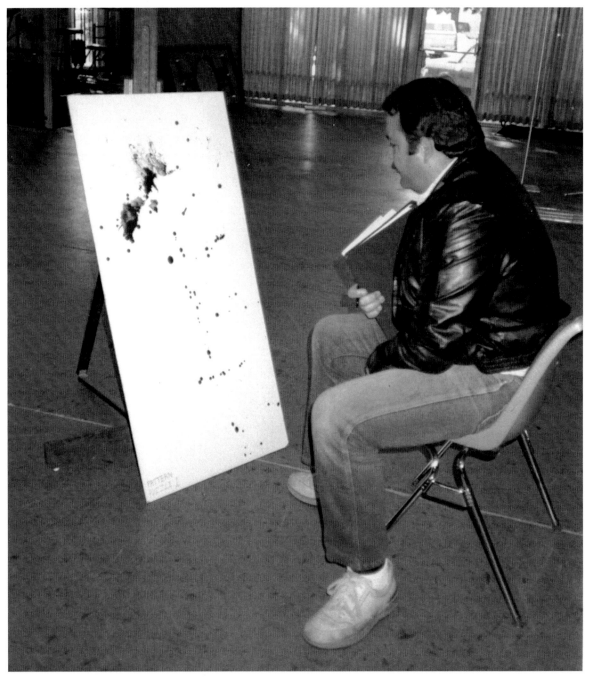

Figure 34-1   Participant pondering the pattern puzzle (floor scenario target).

CHAPTER 34

SPECIAL PROJECTS, PRACTICAL EXAMS, AND MOCK CRIME SCENES

Some very intelligent, capable, and experienced individuals are not good at taking exams. Today's academic systems put heavy emphasis on exam taking and passing to scholastic levels based on the results. Since this seemed unfair in evaluating those who might be the very people who would decide whether or not the evidence was applied in homicide investigations, the class format was expanded to include a variety of ways to make up a satisfactory grade. There is also the objective of determining how much individuals learned as compared with what they already thought they knew when they signed up for the program.

The first item in every class, 40-hour as well as 16- to 24-hour programs, was a pretest. Only 10 questions were asked but they spanned the list of misunderstood principles regarding bloodstain patterns. When the classes began in California, no participant scored all 10 correct, and the majority missed 80 percent or more. At the end of 14 years of training, most respondents were scoring 70 percent or better, and each class had at least one or two with perfect scores. It was also noted that information from each workshop was passed along to future students requiring better and more accurate detail in constructing the course mock crime scenes. Participants of later workshops could not only solve the crime, but they could also point out the errors or deviations from real crimes in how the scenes were staged. Although we were forced to discontinue the program, we did so knowing we had left a good mark on the scientific application of valuable evidence, including some awareness to recognize staged crime scenes.

Each instruction team included someone from a crime lab, someone from police identification, someone from a detective duty station, and someone with a private viewpoint. The class was team taught with instructors from courses in death investigation, homicide, identification, traffic accident investigation, and photography. There was the added benefit of someone who worked in an emergency room stat lab. The information discussed with the various approaches to exercise material was listed on large index cards or plastic sheets for reference in the gym by those acting as an assistant. Feedback was active, ongoing, and extremely beneficial to each program. Following are some of the ways experiments were integrated with the collective experiences from actual casework.

PATTERN PUZZLES

During some of the earlier workshops pattern puzzles were included, like the one being pondered in Figure 34-1. These were sheets of fiberboard or plywood containing a collage that duplicated the overlapping and complex patterns from a case processed by an instructor on the team. The objective was to draw practical oriented individuals into understanding how the information applied to actual casework. The idea was abandoned after a couple of years because the attention of detectives in the classes became glued to the puzzles, and it was difficult to get them interested in the simple exercises. The concept, however, has been very successful in three-day advanced programs presented across the United States.

PHOTO WORK-UPS

These are sometimes included in advanced programs, but not during basics. This is something that can be used if the class is to have limited or no exposure to human blood. The ABC[1] in developing confidence that something seen in a photograph is blood may be included in the training. The drawback is the limit of time. Law enforcement participants may focus on "solving the case" rather than understanding how each bloodstain pattern is classified. A way around this is to include a grouping of photographs with bloodstain pattern evidence (BPE) from several cases, not a single application, and no scenario listed. This requires that the instructor have experience with ample bloodstain pattern casework and that the bloodstains were identified accurately in the cases.

RECONSTRUCTION EXERCISE

No workshop should be complete without some freedom of expression for the participants. Although the transfer patterns offer some innovations, the objective with transfer exercises is to suggest known variation, not allow participants to construct from their own experiences. An exercise that became popular after our first Ohio program was labeled the Time and Sequence experiment. Teams were permitted to list items for their scenario as they developed a theme to reconstruct. This could, if desired, include a case participants were working on at the time or a case that they remembered. The instruction team searched supplies and provided reasonable facsimiles. Some substitution suggestions are listed in Table 34-1.

---

[1]Wonder, *Blood Dynamics,* 11–13. See also Chapter 5, case 1

Item Requested	Item Supplied
Kitchen sink	Small basin, large metal mixing bowl
Commode	Ceramic plant pot with cover (reverse planter with drip dish)
Pornographic video	Advertising video marked with XXXX
Condoms	Office finger cots
Arterial spurt	Syringe barrel spurt
Human arm/skin	Reverse suede leather on wood strip
Human hand	Latex glove filled with sand
Gunshot dispersed impact spatter	High power spring trap
Vehicle (hit and run)	Lawnmower tire
Knife	Magician's retracting blade knife
Window	Picture frame with glass (or plastic)
Firearm	Child's toy gun

Table 34-1

Reconstruction Exercise Substitute Items

The requirements for the exercise were that some sequence of actions was required with a measured waiting time in between. Some of the patterns had to overlap to tie the scene together. At the end teams visited each reconstruction to see if they could quickly identify the steps in construction. All teams became involved in their work and came up with a good array of scenarios (Figure 34-2). This exercise, although entertaining and a good finish to a 40-hour program, also provided valuable training:

Figure 34-2

Drying sequence exercise and reconstruction practice included a victim of torture.

- It taught participants what patterns might look like in a context different from the usual white cardboard.
- It helped emphasize the three-dimensional nature of BPE.
- It illustrated sequences and dynamics of actual crime events.
- It brought together review of several pattern categories in one overall context.
- It encouraged team problem solving using bloodstain pattern analysis.

SPATTER PRACTICE

This is essential to any training where the participants plan to use some form of reconstruction of the origin. In Chapter 4, the logic and errors in logic with origin reconstruction were discussed. The first emphasis is on proper training in how to select and measure blood spatters. National "Teach the Teacher" programs advocate learning a subject three days in a row to lock it into permanent memory. In reconstruction of the origin, three types of practice are used to commit the information for participant future use:

1. Measurement exercises often are developed with drip techniques at various angles. The behavior of large blood drops in free fall onto a slanted, cardboard target do not accurately reproduce what will be examined and measured at a crime scene with some kind of impact spatter pattern. Actual impact spatters, especially from gunshot, may be small and require magnification to see well enough to measure. As an alternative to using the large drip stains to learn to measure, a selection of angled spatters from known impact angles were cut out, pasted on a single sheet of paper, and photocopy-enlarged eight times. The cut offs for these stains was listed and used to show participant cut-off points on actual impact spatters.

   Next, a sheet of 25 spatters enlarged two times was given as an assignment for participants to measure. Any person having trouble was given personal instruction regarding cut-off points for measurements.

2. String reconstruction practice can be included various ways. One method has been to angle a sheet of plywood or masonite and drip blood at various angles as noted on a carpenter's level. This is associated with drips at angles and familiar to participants who have learned to measure with that technique. This was found to confuse participants since drips are never measured at crime scenes.

   Another way to approach training is to provide a 1:1 scale photograph of a surface from a known impact event. The California workshop included a photograph of a textured wall adjacent to a beating. The picture can be mounted on a plywood sheet or package carton cardboard, then covered with a Plexiglas sheet (available at hardware stores everywhere) and attached to any wall with duct tape (wall paint may be removed from the wall with the tape, so inquire first if this is acceptable). An example of an exhibit used in traveling workshops is shown in Figure 34-3. Water markers should be used to circle stains and then be wiped clean between teams.

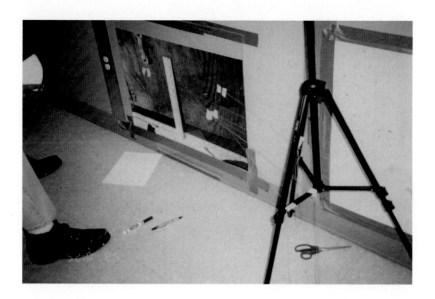

Figure 34-3
Actual crime scene photo used for string reconstruction practice.

This is provided as additional training in selecting the proper stains to measure and practice (not graded) in selecting spatters from a single event, locating the area of convergence, testing the assumption that the spatters were from one impact, and providing practice in measuring stains of the relevant size and shape as a real crime might involve.

3. A mock crime scene should be mandatory for every basic workshop taught. Problems with mathematics applications, differences between theoretical and actual behavior of blood require confirmation that the correct approach to bloodstain pattern interpretation is demonstrated to participants. The mock crime scene is an excellent medium to achieve this. It is also a good way to focus participants on a reasonable facsimile of their future workups. The scene may be constructed in a place as little as a corner of a room or in a specially manufactured room with plywood or heavy cardboard walls. The size is also variable but an 8-foot cube worked well in the California programs. A law enforcement evidence garage bay was used in one road show and a men's bathroom in a college setting was used at another venue. A scenario is worked out, usually based on previous casework. The mock crime scenes have also been beneficial in reexamining cases over a long period of time for academic purposes.

Each scenario should include at least one example of each of the six major pattern groups: impact spatters, cast offs, arterial damage, transfers, PABS/dry, and volume. A note regarding cast offs—they need not come from a weapon during a swing. The pattern can be used as originating from insignificant actions such as a bloodied object falling to the floor and casting drops at an adjacent wall, or flicks from removal of the victim by coroners or paramedics.

Only one impact event should be included. If multiples are necessary they should be separated so that complex patterns are avoided for a basic class. Participants should be cautioned not to draw all strings from a single surface area of convergence or to conclude a point origin.

PRACTICAL EXAMS

Practical exams sometimes are considered time consuming and trouble to manufacture and store. They do, however, provide an index of how participants are assimilating information from both lecture and laboratory exercises. Nothing is more disappointing to a concerned instructor than to find out on the final exam that participants have missed the whole objective of the program. This is especially true when conflicting information has influenced expectations in regard to bloodspatters. Although law enforcement tends to standardize programs to be interchangeable and consistent, BPE benefits from variation. No single 40-hour course can provide all the information possible. Attendance in more than one program is not duplicity, but actually advisable. Each program has something to offer that other programs may not. For this reason those attending lectures by different instructors may be faced with shifting focuses. The objective should be to learn from all. Practical exams can identify confusions before the end of the program.

Two areas both lend themselves to and require confirmation of training with practical exams: spatter (spots, bloodstains) identification and clothing examination.

Spatter classification exams should be included somewhere in a program. In the California courses the material was presented as a practical exam on the third morning of instruction. This gave three more days of program to correct any misconceptions or confusion that occurred as a result of the first two days. Two plasticized, white cardboard sheets were given to each team with the instruction to rotate so that all teams had the opportunity to examine all exhibit sets. A grade was assigned to the first set analyzed, and based on recognition of the type of pattern (impact, cast off, arterial), further classification (gunshot, beating, swing, drip, cessation, spurt, gush, blood into blood, exhalation/respiratory); and how the two sheets were positioned at the moment of the event distributing spatters. When the answers were given, the logic involved in interpretation was pointed out, which evolved into a training session in itself.

Clothing exam was added after observations that law enforcement in California were not consistently trained in clothing collection. This was a split between departments where a separate coroner's office collected the victim and the clothing. The approach was to place the victim still dressed into a body bag, transport to the morgue, and the clothing would be removed when the autopsy was performed. The logic was that trace evidence items would be lost at the scene if the clothing was removed there. Unfortunately the bloodstain patterns on the victim were often lost from contamination of blood seepage within the body bag on the way to the morgue. At some point after adding this section to the general bloodstain pattern training program some jurisdictions changed their protocol in body handling. In those areas the victim was thoroughly photographed in place with and without markers, then lifted and placed on top of a sheet or drop cloth. The clothing was removed over the sheet and allowed to dry. Care was taken to not transfer blood. Each dried article was bagged over the sheet/drop cloth. The nude body was again photographed, placed in a bag and transported for autopsy. The clothing went directly to the crime lab or evidence storage.

For economy of time and space to conduct this practical, each set of clothes was constructed like a mock crime scene and then placed in set numbered shopping paper bags. Each set included a pertinent question to be answered. The grade was on recognition of patterns and answering the question correctly. Some examples of sets used follows.

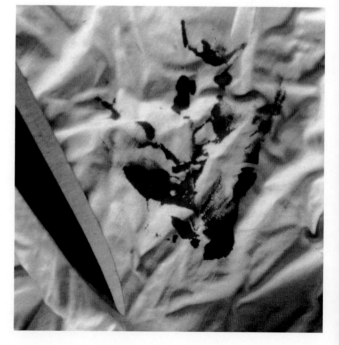

Figure 34-4

Towel with knife wipe but also with palm print of person doing the wiping!

1. A transfer on a towel was given with a set of knives collected at the scene. The question was, which knife was used in the alleged attack and how did they arrive at that conclusion (Figure 34-4).

    An interesting note was that in addition to the knife imprint, the palm print of the person who wiped the knife is also recorded.

2. An exhibit containing imprints of an assailant's shoe was given out (Figure 34-5). Then a set of rubber sole sandals, all different styles, and exemplars for the shoes were given. When the team identified the correct style of shoe it was pointed out as a class characteristic because the imperfections were not consistent with the print, only the main design of the sole (Figure 34-6). A set of six sandals, all the same type but different colors, was then given to the team with a set of exemplars showing patterns for the various individual shoes. The question was, which shoe made the mark at the scene and how did they arrive at that conclusion (Figure 34-7)?

Figure 34-5
Crime scene exhibit.

3. A shirt and a pair of jeans were presented with spatters on the shirt and a transfer on the jeans. The items were from an assault situation. Two stories were given regarding how the patterns occurred and the team was asked to decide which was consistent with the patterns or to present a third story.
4. A set of clothing was presented where a suspect had cradled the body of an acquaintance while an artery spurted. The question asked was relative to the wearer of the clothes being involved in a crime or receiving the blood spatters from noncriminal intervention.

Figure 34-6
Shoes with the same class characteristics.

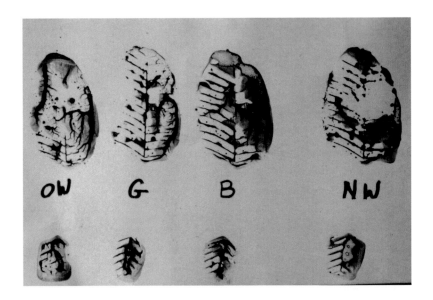

5. A shirt was presented with cast offs over the shoulder and on to the back. The question asked was relative to characteristics of the wearer of the shirt and action necessary to distribute the pattern recognized.

6. Two pairs of jeans with spatters on the legs were submitted, in which one set was worn by someone kicking the victim and one set was worn by some one standing next to the victim who was being kicked. The question was, which of the pants were being worn by the probable assailant?

DEMONSTRATIONS

The practical exams are demonstrations in themselves but a number of other exhibits have helped maintain attention from participants and provide additional exposure to items that may be seen at crime scenes. These include examples of flyspeck distributed blood-stains, unique transfer patterns, and respiratory impacts such as wheeze[2] and GDIS (Figures 1-9 and 1-10). The smoke detector in Chapter 17 was given to the author. The exhibit was fitted within a plastic box and can be passed around in a class without fear of biohazards. Transfer patterns on material that might be found at crime scenes such as carpets and painted walls can be encased in plastic boxes for demos. Demonstrations showing how attackers leave blood prints in the commission of assaults are entertaining as well as good at leaving a visual memory. Induced sneeze and coughs have been demonstrated in this same context. Blood drawn from the illustrator into EDTA can be used on their own skin and in their nose or mouth. EDTA is a food preservative and not poisonous.

During early workshops the class was transported to firing ranges for gunshot experiments. Although these are fun, what was learned did not compensate for the time and trouble to arrange. In fact many participants of early classes became focused on GDIS and ignored the many other important lessons. So the exercise was deleted but exhibits were provided for the class. One was a collection of pig skin suede entrance and exit wounds made from various handguns available. The smooth side of pigskin suede looks and behaves like human skin and records bullet passage much like the real thing (Figure 34-8).

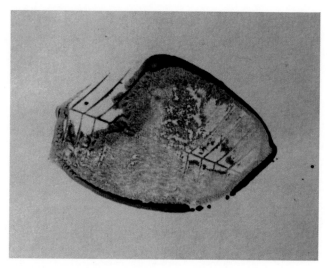

Figure 34-7

One shoe from the exemplar made the print in the exhibit. A second exemplar is given to show how taking exemplars is important too.

---

[2]Wonder, *Blood Dynamics*, 49.

Figure 34-8

Pigskin suede smooth side used on blocks of agar gel to duplicate the appearance of entrance and exit wounds for various handguns.

FINAL EXAMS

Final exams are varied in this type of training. The emphasis on three-dimensional context of the crime scene is presented in some, and preferable to the standard multiple choice format. The drawback in courses where the only grade is based on the final exam is that many extremely capable investigators are not good at taking written exams. In this regard using practical exams along the way and making the mock crime scene a major part of the grade assists those who might appear to do poorly based on a single final. Over the years some questions were found to be misunderstood and fail the objectives. These were replaced, and good questions were retained. As a rule those questions requiring complex thinking did not go over well in a limited time period of the exam.

The most important point in exams was "read the question carefully." Many people speed through an exam and make mistakes simply because they do not answer the question asked. Bloodstain pattern evidence is a thought-provoking evidence. The rewards are great but to benefit one must apply a greater degree of logic and sequencing than for other physical evidence. For some it is just too much trouble within the hour or so for a final exam. What those who give up early because of this do not find out is that once you get used to the logic and understand the dynamics it can come naturally without all the hard efforts. Unfortunately the beginning stages require work. Reading an exam carefully as if one were responding to questions from opposite counsel on the witness stand is a good beginning. Many participants from the various workshops have felt considerably different weeks or months after the course. This is another reason to break up the final grade so that the stress of an exam does not influence the benefit graduates appreciate.

Much information is available from bloodstain patterns, with additional knowledge shared and acquired from participants attending workshops. Some of the case presentations represent many years of discovery for the identifications and interpretations of patterns. Course participants may end up feeling they can provide all the same information in a brief view of a crime scene after finishing the course. The author believes final exams should be difficult to help emphasize that one 40-hour workshop does not make an instant expert. After the workshop, experience and practice are necessary to acquire expertise in the field.

THE FUTURE OF BLOODSTAIN PATTERN EVIDENCE

Figure 35-1 Science Magazine, *reprinted by permission of the American Academy for the Advancement of Science, shows a common artist plewd representing blood drops. The first thing usually taught in bloodstain pattern workshops is that the plewd is not an accurate depiction of blood behavior. Still, misconceptions exist between symbols and facts, as this illustration shows.*

CHAPTER 35

RESEARCH IN BLOODSTAIN PATTERN EVIDENCE

There are three primary ways that research can be initiated for the benefit of bloodstain pattern evidence: workshop exercises, case reconstruction, and academic projects. Each of these can provide observations to study in detail later. As direct research there are advantages and disadvantages to each approach.

OVERVIEW

Bloodstain Pattern workshops are not normally good sources of research data although observations during the programs can provide problems to be studied at some other time. Two advantages to workshop formats that benefit research are the availability of human blood and the format to repeat exercises in multiples. A factor in research is to repeat the experiment enough times to determine if the results were random aberrations or a true consequence of the parameters of an individual test. This can be achieved when specific parameters are assigned to the number of teams, usually four to six per workshop, participating in repeated programs. Human blood is not easy to obtain, therefore having a training program where the substance is available provides an opportunity for study. Workshops, especially those presented annually or more can provide repeats sufficient to compile data of research using human blood in a long-term study.

Unfortunately workshops normally are not designed to study observations. The experiments are usually developed as illustrations of known results followed by case presentations in lecture. If any experiment is repeated enough times an unusual result will occur. When this happens in a workshop environment it is usually ignored, not recorded or studied further. Programs do not provide enough time to explore aberrations, and explaining unexpected results may be viewed as disrupting to the flow of lectures. A few situations exist where the objective of the program is for graduates to contact the instructor later rather than apply their own knowledge. In these, training is usually incomplete, omitting tricks of the trade techniques. As such, the question of whether something is an aberration to expected experimental results or an unexplained common occurrence may be left unresolved.

An example of the unexpected in a workshop occurred with the spring trap device example of impact. One participant pulled the pin on the device and projected blood spatters in a recognizable pattern up to the 12-foot-high ceiling. No one else had ever done this in 25 workshops with at least 18 releases of the trap per program. The man was challenged that he couldn't do it again, and he proceeded to do it. None of the instructors could explain how this one person could project spatters from a level spring trap over 9 feet up into the air. The trap did not leave the table. Exceptions occur and they may even be reproducible but they do not and should not be used to define the event. These qualify for research only after they have been repeated adequately under control conditions for verification. When exceptions happen in a workshop they are usually explained based on educated guesses and ignored.

Observations that deviate from the expected, however, can provide subjects for research later. One such observation was made during workshops regarding the overall appearance of bloodstains from packed red blood cell transfusion pouches as compared to spatters seen at actual crime scenes.[1] This was mentioned in the chapters regarding blood products used in workshops. Much more research into the effects of red blood cell concentration with respect to dynamic acts would be beneficial to future investigations, especially since red cell concentration affects non-Newtonian behavior. Subjecting different known hematocrits (red cell ratios) to the routine experiments of a workshop with all data recorded could serve as a research project.

Reconstruction of casework as a source of research material may be valid if the reconstruction is conducted without a preconceived result (i.e., bias). Two big advantages to using casework reconstruction as opportunities for research are the association with actual crime scenes and the authorization and funding to do the project. The major drawbacks are that the end isn't always open to recording all results, and time constraints usually prevent repeating the experiment enough times to provide frequency data. A format to reconstruction should also vary parameters so that the conclusions are tested. A way to do this is to repeat the experiment as the suspect claims, the way law enforcement officers feel happened, and one or more alternative ways that might be at variance with the other two.

Unfortunately this is not always the way reconstruction is carried out. A story was related to the author during a conference, regarding an alleged reconstruction that should not qualify as research. The apparent objective was to illustrate that an event claimed by the accused could not happen as stated.

---

[1] Personal conversations with Terry Laber and Barton P. Epstein during and following their annual blood-spatter workshop in Minnesota.

When the results did follow what the accused claimed, the designer of the experiment stated on video tape, "That's not right. We have to do it again." In this kind of experimentation, results not desired were unlikely to be shown in court or recorded for future reference. Bias is not acceptable in research data.

An example of casework application of research should also include observations and recording of a number of criteria for results. In the past, law enforcement has approached reconstruction by comparing only where spots of similar sizes are found on surfaces suggestive of the scene. Far more important is at least recording directions of travel and stain distribution, size and spatter density. For quality research the results must be examined for *position* and *pattern* relative to the dynamic blood distributing event. Such experiments are good for showing where blockage and absence transfer patterns would occur during the commission of a crime, and how blood drops were distributed in an array. A study that would greatly benefit case applications of bloodstain pattern evidence (BPE) might be a formal comparison of assault, CPR, and respiratory distributions, reconstructed and compared with actual paramedic and emergency room staff work.

The major limitation from reconstruction research is the lack of time to repeat experiments for reproducibility. This makes recording and reporting of case results for comparison with future and past cases much more important than for workshops or academic research projects. The case sharing format of IABPA annual meetings provides opportunities to compare results and increase statistical data. Unfortunately the entire meeting program is not yet published for sharing with the entire membership. Attendance at meetings is required for the case content. To date, no recorded statistical records of the presentations have been shared with members.

Academic sources for research projects have not been common in the past but fortunately are gaining interest at the present time. Original research projects require considerably more work and an independent source of funds as compared to projects conducted during workshops or for specific case reconstruction. Safe blood must be purchased and parameters of the experiments controlled with location of the experiment sealed off from other scholastic activities. The benefit of this source of research is the academic approach to the projects with specific standards of study outlined. Professors are usually excellent in keeping the study free of bias and open to any results that occur. A serious drawback is the lack of actual crime scene experience. Interviewing an experienced investigator, having an advisor in the field, or an apprentice program is an ideal addition to academic research projects. Fortunately, some college professors in forensic science have also worked as experts in actual casework.

INITIATING AN ACADEMIC PROJECT

Here for review and consideration are some suggestions for approaching the initiation of an academic project:

1. Select topic area.
2. Perform literature search and define project.
3. List data to be recorded, variables to be studied, and controls to be set.
4. Find and arrange secure area for project.
5. Design and manufacture devices for reproducibility.
6. Obtain funds for project.
7. Conduct study.
8. Evaluate data and draw conclusions.

SELECTING SUBJECTS FOR RESEARCH

True research objectives begin with a problem or question rather than a preconceived objective or result. The problem can follow observations during a case or workshop experiment. Whether the answer to a question will benefit the future of the evidence or not may require experience with crime scenes or discussion with individuals who have case experience. Reviewing observations in casework will help recognize the possibility of individualizing features. Just for interest, here are some of the projects considered by the author as studies to contribute to investigative use of BPE:

1. What are the nature of and conditions for displacement cast offs? This comes from interpretation of Sgt. Reichenberg's TAI case (refer to Chapter 26), but would be interesting to study from a completely independent viewpoint.
2. What effects do hovering aircraft have on wet blood on the ground or on suspects? Such information could apply to highly publicized cases, caught in the act homicides with news coverage, and under strong wind weather conditions. This was a question in Dr. Davis' DNA case (refer to Chapter 28).
3. How does close contact alter patterns such as impact spatter and arterial? This question comes from recognition of GDIS on a pillowcase (Figure 35-2) used to prevent spatters (back spatters) distributing on the shooter. The identification of the GDIS pattern was from area of convergence, alignment, and distribution of stains. This is the pillow on top of the woman's head in Figure 19-8.
4. What causes color changes in bloodstains? We know red blood can be arterial but it can also be from carbon monoxide poisoning and some chemical poisons. Chances are that exposure to freezing, extreme heat, road debris in the form of battery acids, and road salts can affect color also, but not all of these have been subjects

of formal study to date. Does blood change color in a predictable manner with age as stains dry? This latter would be an excellent study for the commendable identification technician photographers involved with the evidence.

5. How does velocity within the major categories affect distributed bloodspatter patterns? The present velocity scheme involves set ranges for velocity such as 100 feet per second or greater for gunshot and 20 feet per second for beating. Baseball players can swing bats faster than 20 feet per second and fly wheels can cast off at 100 feet per second. So what are the variances of velocity within each of the major events: cast off, impact, and arterial distribution?

6. What effects does blood of different cell ratios contribute to dynamic events? Cohesion and adhesion are different, thus variance should be expected with surfaces that absorb liquid but not cells. Comparisons of cast offs from unfinished wood, finished wood, and metal could serve as a project.

7. How does cohesion affect the traditional exercises in bloodstain pattern workshops? We know that blood has different ratios of red cells, which can change the behavior from Newtonian to non-Newtonian or back. The behaviors are presently too technical and complex for the usual bloodstain pattern workshop. Using blood of different recorded hematocrits in research experiments can provide information for future incorporation of this information into training and case applications.

8. What difference does the speed of a drop of blood make on measurements? Much more needs to be done with this given modern photography and measuring devices. There is no doubt that the difference in drop speeds affects the shape of a bloodstain but there may be formula, variables that can be used to adjust and modify so the information follows logic.

9. How can we use rheological streaming in the identification of arterial damage patterns, especially carotid arterial breach or severance?

10. What effect does cloth thread count have on the interpretation of stage of coagulation?

11. How can we define and apply information from blockage patterns and areas of convergence to confirm and independently corroborate reconstruction of origin techniques, string, laser, and computer applications? Formats and schemes to use these with other factors should continue to be studied.

12. Can the properties of convex meniscus, reflective bending, and/or thixotropic friction be used in a device to measure non-Newtonian behavior of freshly shed blood?

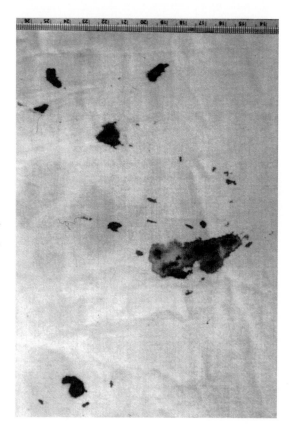

Figure 35-2

Pillowcase on top of victim's head when shot fired postmortem into back of head.

This is far from an exhaustive list of scientific questions that could benefit crime scene investigation. Studies regarding the distance a blood drop falls relative to the diameter or alternative means based upon a standard blood composition are attempts to reinvent a square wheel. They work well in a laboratory, require little exercise in logic, and write up well for a paper. The conditions for application do not exist at actual crime scenes. Time and funds are wasted in such studies.

LITERATURE SEARCH

Academic research should always begin with a literature search. The purpose of a literature search is to define your project as well as limit it to manageable parameters (variables). It will help you decide what data you need to collect. Much has been written regarding directions of travel, velocities, and arrangement of spatters. For future information, literature explaining non-Newtonian drop separation should be included and medical references could be reviewed for injury characteristics that may contribute to the amount of blood as well as manner of distribution. Most important is that you can use the information from other studies to economize and concentrate your own project.

A literature review starts with a list of keywords. This is a format developed by various organizations such as the American Academy for Forensic Sciences, where all abstracts are to include a list of keywords upon submission. The objective is that further research will follow science papers and the keywords may assist the preliminary literature search. Some uncommon keywords and literature sources for BPE may be added:

Aerodynamics
Biorheology/Hemorrheology/Rheology
Cohesion
Colloidal chemistry
Hematology (Hematocrit, Hemastasis, Coagulation, and Sedimentation rate)
Medical texts for Artery, Hypervolemia, Vascular flow and injury
Newtonian fluid mechanics
Non-Newtonian fluid mechanics
Physics projection formula

An essential reason for including medical keywords and thus medical research literature in the search is because blood has been the focus of considerable research in that area. Medical research has considerably more funding than forensic science, let alone BPE. There is a financial benefit from using medical references if they apply to a project. It is advisable to actually review the references listed in literature. Interpretation is essential in forensic science and that includes each author's viewpoint of reference material. Occasionally references are checked, and it is found that they do not provide the information attributed to them. The conclusion follows that an author quoted the reference from another author's bibliography without actually checking for themselves.

LISTING DATA TO BE RECORDED, VARIABLES TO BE STUDIED, AND CONTROLS TO BE SET

After the literature search, a good next step is to list the variables that can be controlled and how. Ideally all variables except one are controlled, or fixed. The single parameter is then varied against a time scale to provide information regarding that one condition. For instance if drying time were being studied in a research program, the source of blood would come from one donor, lighting conditions would be uniform, angle of blood flows would be the same, air/breeze conditions and humidity would be controlled, and amount of blood distributed would be set. The parameter varied could be temperature. Time intervals would be established and test methods such as blotting, rubbing, and changing direction of tilt would then be applied at different temperature settings versus time. If the project were conducted outside, additional variables might contribute to humidity, light, and wind conditions.

One area that is lacking in most published results of experiments with blood is the composition of the blood used. The concentration of red cells in blood samples affects the behavior of blood in several ways—coagulation, sedimentation, diffusion and aspiration, flow, spread after being distributed through air to a recording surface, and the size of a drop upon separation from a blood source. For the benefit of future research it is essential that records include the nature of the blood source used in any research or reconstructive experiment. The minimum information should include:

- Donor (human, animal, substitute)
- Type of collection (whole blood, packed cells, reconditioned blood products)
- Hematocrit (ratio of red cells in material used)
- Age of material (out-of-date blood bank, fresh within 72 hours, freshly drawn)
- Anticoagulant used

Recording time intervals, temperatures, and air current conditions are good for future reference whether needed in the study or not. Blood is both in limited supply and difficult to obtain due to discoveries of disease processes within the past 20 years. What was simple and easy to perform 25 years ago is no longer possible with readily available materials. It is essential, therefore, that data be recorded so that the next generation of researchers need not repeat all the same preliminary experiments.

FINDING AND ARRANGING A SECURE AREA FOR THE PROJECT

Workshops and case reconstructions usually have specific areas that may be used in hazardous conditions (i.e., with gunshots and biological body fluids). The academic environment does not provide for such on a routine basis. Facilities are routinely used for classes and studies that do not involve a need to protect and isolate. Controlled environment rooms are costly and may be dedicated to specific grants from companies and industries that would benefit from the projects. Bloodstain pattern exhibits require a large area where they can be left to dry undisturbed. Surrounding surfaces then need to be bleached clean after the study is completed. Specific consideration should be given for this before starting a research project.

Figure 35-3

"Hot Lips" device fashioned to reproduce cough-type events.

DESIGNING AND MANUFACTURE OF DEVICES FOR REPRODUCIBILITY

Research techniques require repeating experiments varying a single or a small group of parameters to study reproducibility. With BPE this means recreating impacts, cast offs, and arterial pressure of the same levels of force for comparisons. In workshop conditions variation isn't a problem. The simplest procedure is to allow students in protective clothing to be as creative as they like. This is also a reason why workshops are not ideal venues for research. Some devices that are used in the workshops, however, can also be used in research. These include the spring traps, free-fall devices, supports for angular drops, and arterial pumps. The author has had the benefit of some other devices using them in class, casework, and research.

Devices can be simple as with the "Hot Lips" system, shown in Figure 35-3. This was developed to recreate bloody coughs such as shown in Figure 35-4. Note the variation of sizes, and that they reoccur throughout the whole arrangement. The same arrangement of size variation can indicate a reproducible distribution of extreme size variation. If seen at a crime scene, the pattern might vary simply by the amount of blood and mucus in each cough. An experiment could check this by varying the levels of each in the pump bottle. Egg white may be substituted for mucus. The levels of fluid in the bottle should be near the tip end of the aspiration tube to create the effects of cough drawing air into the action. Filling the bottle with blood and egg white will simply clog the system. Marbles can be added to help break up the egg white and add air to the mixture.

Slow Bullet was mentioned in the workshop experiments (Figure 35-5). This provides interesting contrasts to other techniques in reproducing GDIS (gunshot distributed impact spatter). Note the mist from an artist spray paint device (Figure 35-6). Compare this with the pattern from Slow Bullet in Figure 35-7. Other dynamics that could create patterns with the same or similar size ranges of spatters include sneezing, wheezing, and initial breach of an artery under very high pressure.

A series of experiments for the purpose of studying traffic accident investigation problems illustrated that obtaining vehicles, gauging speeds, etc. was not easy and trouble free. A drum device was manufactured to test various questions regarding impacts at given rates of speed by motor vehicles (Figures 35-8 and 35-9, see page 337). Speeds were not freeway caliber but varied within those questioned for in-town driving where pedestrians would be encountered. Refer to Figures 35-10, 35-11, and 35-12 (see page 338).

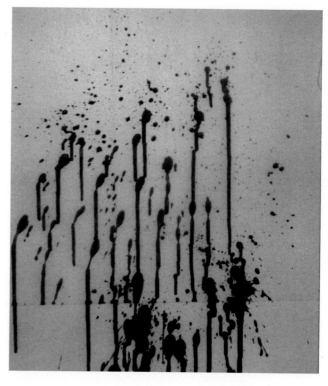

Figure 35-4

A simulated bloody cough produced with "Hot Lips."

Figure 35-5

The trap device nicknamed Slow Bullet has been used to create reproducible GDIS where firearms are not allowed or would create unnecessary hazards.

Figure 35-6

Artist spray paint assembly used to literally mist the target.

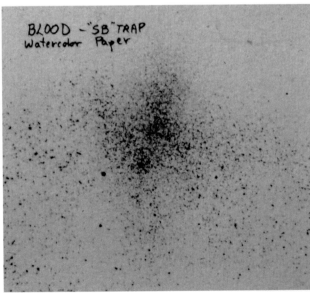

Figure 35-7

Impact from Slow Bullet to compare with the artist spray device.

This device helped when we wanted to know the exact speed at which an event occurred. Of course some comparative studies can't be completed without the real dynamics involving a motor vehicle. A TAI question that may be important to law enforcement and emergency response vehicles is whether or not a splash pattern was created by a vehicle hitting a downed pedestrian or whether the vehicle hit a blood volume in the roadway. This was conducted at a workshop in which a three-quarter ton, full-sized pick-up drove at about 20 mph over a plastic pouch contained unit of blood (Figure 35-13, see page 339). The same truck drove at the same speed over the blood volume (pool) exposed on the roadway (Figure 35-14, see page 339). The latter event was less contained and distributed blood over a wider more random area (Figure 35-15, see page 339). Viewed from directions of travel back toward the center, the area of convergence for the splash from the volume would be the width of the pool, not the compact size of the victim, unit pouch. This is because blood distribution was from the circumference of the volume stain, not from the center as with the explosive rupture of the blood pouch.

A device that should be a part of routine crime labs is the rotating firearms target (Figure 35-16, see page 340). This was used in a number of exercises, and then donated to a law enforcement agency. The separate parts of a gunshot may be recorded on the spinning target as in Figure 35-17 (see page 340).[2]

A device that worked well for a case study, and could be used in pure research, is the reproducible pivot. The question asked for the project was, "what happens to bloodspatter distribution at and immediately following a gunshot to the face of a victim?" A possible answer from the crime scene photos was that the tilting of the head immediately following a gunshot was responsible for the specific distribution of spatters on the victim's face. The pivot device was used in the case presented in

[2]Wonder, *Blood Dynamics*, 131.

Figure 35-8

Drum device for testing events at various velocities.

Chapter 11 (Figures 11-15 and 11-16). A toothpick held the device in place until the moment of the gunshot. When the bullet was fired the toothpick released and the head pivoted in a reproducible manner. It was shown that spatters could be recorded on the ear and eyelid when the head pivoted at the moment of the shot.

Another note regarding research may have additional applications besides BPE: blood and water effects (PABS/mix-water). The author's interest in the effects of trying to wash away bloodstains began with the master's thesis project. Although it was only a minor note with the thesis, it is now a much more applicable subject. The case presented in Chapter 7 involved bloodstains in the kitchen sink. Because the timeline was important between acquiring a bloody nose and the gunshot wound to the head (i.e., supported with witness statements), what

Figure 35-9

Drum device for reproducibility of events against a moving linear target.

Figure 35-10
Low speed, around 6 mph.

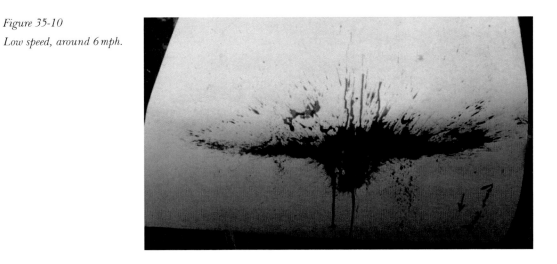

Figure 35-11
Higher speed, around 10 mph.

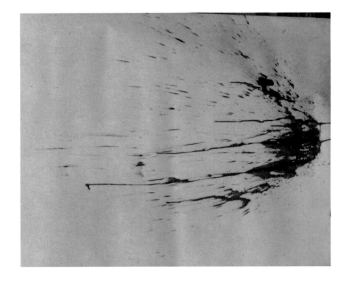

Figure 35-12
Maximum speed for the drums was around 12 mph.

Figure 35-13

Truck impact with a blood bank plastic pouch, whole blood, hematocrit 48 percent.

Figure 35-14

Truck impact with volume (pool) of blood from Figure 35-12.

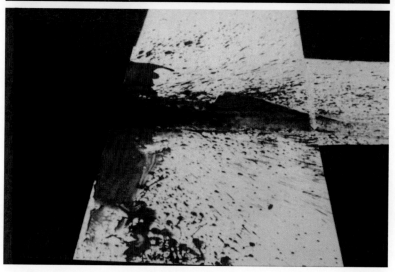

Figure 35-15

Closer view of splash from truck hitting pooled blood.

action caused the stains in the sink may have been of significance to the investigation. Three possibilities were considered:

- The sink had water spots when blood was spattered into it (Figure 35-18)
- There was blood spattered in the sink, then water was flicked on to the spatters (Figure 35-19)
- Someone with bloody hands held an ice block and threw it into the sink (Figure 35-20)

The variety of colors and uniformity of pigment in each stain of the crime scene photograph seemed to suggest the third action as the most likely one (Figure 35-21). This was confirmed by the fact that a piece of the tile counter top was broken off and laying on the floor (i.e., the block of ice was hit against the counter before or after being dropped in the sink).

Additional experiments were carried out with blood and water mixed before or after depositing on fabric. This was mentioned in *Blood Dynamics*.[3] Various pattern notes were observed, but the study needs to be repeated

Figure 35-16
Rotating target device.

Figure 35-17
View of target spinning at about 25 mph when gun fired.

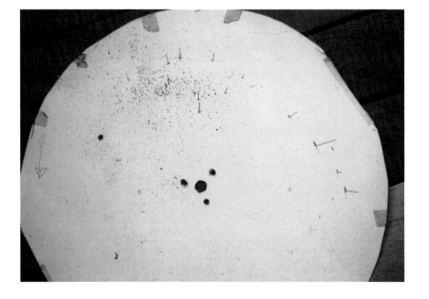

---

[3]Wonder, *Blood Dynamics,* 115.

Figure 35-18

A wet sink had blood spattered on it.

before it is a valid research result. The interesting observation may have application in situations where the crime scene has been cleaned with just water. Figures 35-22 and 35-23 show various patterns for blood on dry fabric, wet fabric, and dry then washed out. The last stain was washed until the center tested negative for blood by o-toluidine (Clinical Lab Science test for occult blood). The center was cut out and emerged in physiological saline at 37 degrees for

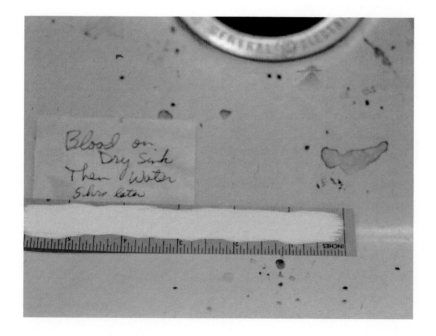

Figure 35-19

A blood-spattered sink had water flicked on it.

two hours, then the button was used in an immuno-aggregation procedure with typing cells. The original blood type was determined. The reason this was possible is that blood dried on the fabric. Washing washed away the hemoglobin, but left the cell stroma (enveloping "skin" of the cell). The cell stroma contains the blood serotypes, which could still be used to identify blood types. Whether this would work with DNA determinations on the nuclear material of dried leucocytes within a blood sample could be studied by the appropriate specialists.

Figure 35-20

A half-gallon milk carton sized block of ice was held in bloody hands and thrown toward the sink.

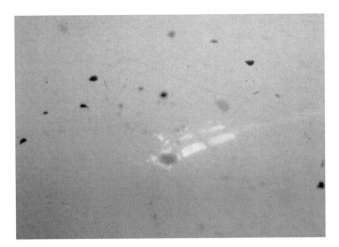

OBTAINING FUNDS FOR A PROJECT

It has been pointed out periodically that funding for bloodstain pattern research is near, if not, nonexistent. No chemical or instrumental manufacturer would benefit from providing funds. The only sources are academic, casework, and out of pocket oriented. This has had an effect on the amount and types of research conducted. In February 2006 during the annual meeting of the American Academy of Forensic Sciences, a committee of five individuals with varied backgrounds in bloodstain pattern analysis met to discuss a funding grant for research in BPE. Representatives were from Massachusetts, Florida, Ohio, California, and Alaska.

The fund would be managed privately but ultimately be turned over to control by the AAFS. Further discussions will be held at the 2008 Washington, D.C. meeting. Future input may be sent to the author care of the AAFS. Initial grants would be in the range of $200 to $300, expanding to up to $1000 in five years, and may be posted at AAFS meetings. Submissions require a literature search (including references from other science fields as well as forensics), statement of benefit to actual casework, a list of parameters to be studied with the projected context in which the experiment will study them, and timeline for completion and progress updates. Terms and applications may be modified depending upon input and further discussions. Anyone interested in submitting ideas, comments, or requesting a grant should contact either the author or G. Michele Yezzo via the AAFS office.

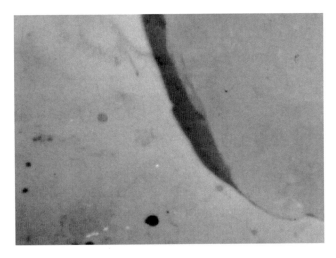

Figure 35-21

The stain color and uniformity of the actual evidence.

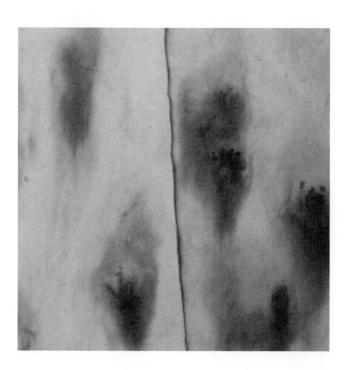

Figure 35-22

Blood on dry fabric and washed with just water.

Figure 35-23

Blood dried and washed until the centers were negative for blood. The clear centers gave blood types (B Rh+) with an indirect agglutination test.

CONDUCTING STUDY

If all the previous details were noted and resolved for a study, the actual performance will be smooth and rewarding. Something always goes against expectations, however. When this happens, taking notes and describing as much of the conditions at the time of occurrence as possible helps to establish why and what the chances for reproducibility might be. Not all data collected will be of use immediately but collecting it will save time and funds in the future.

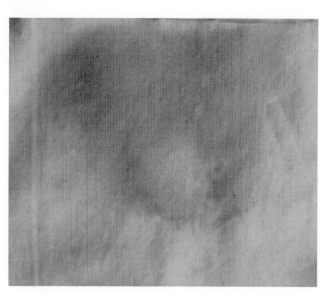

EVALUATING DATA AND DRAWING CONCLUSIONS

Understanding the true benefits of any research project takes time and should involve the use of statistical tools. Computerization has greatly improved the final analysis of research projects, and must be considered in any evaluation process. The most significant benefit of computer application is in eliminating bias. The success of eliminating bias, however, will depend upon the program and the data input being nonbiased.

Figure 36-1 Anita Y. Wonder in her Sacramento office.

CHAPTER 36

SUMMATION

IS THIS ALL THAT IS AVAILABLE?

No, not even close. In going through slides, photographs, case reports, notes from literature searches, and correspondence with individuals from many different science perspectives, it became clear that the effort to put all 25 years of experience with bloodstain pattern evidence (BPE) for only one individual in a book fell short of the mark. Someone once told the author to not teach everything she knew. Relating this later has brought smiles to the faces of students and associates. The moral is that what any of us know today will be less than what we can know tomorrow. Teaching all at one point in time leads to incentives to learn more so that it becomes impossible to teach everything for all time.

The acquisition of new knowledge is exciting and drives us to discover. On the other hand finding that what we believed 25 years ago, and have consistently shared with others since, has completely changed with discoveries from fields totally separate from our own, can be depressing and exhausting. The solution, however, isn't to ignore or deny but to pass on where we suspect changes are indicated, and let a new generation of analytical minds tackle the new perspectives of the evidence. It has been the objective of this work to point out how BPE is a science and can be applied in an objective approach. Some changes must be recognized. It isn't a matter of whether we wish, and/or vote, to accept them but rather a matter of when are we going to acknowledge that science changes and so, in time, must bloodstain pattern analysis. The question is when will the changes take effect.

The present time is probably the most important period for this evidence as a science discipline. Reviewing information from as widely separated areas as Belgravia, Sydney, Bogota, Hong Kong, Chicago, and Toronto suggests that much needs to be done in correcting misconceptions and misapplications. One of the most surprising statements that has been encountered recently is that experienced, capable law enforcement officers have never seen cast off bloodstains at the scenes of crime. From the author's experience and logic, cast offs are among the most common of bloodstain patterns, more so perhaps than impact spatters. If individuals capable of seeing, and actually trained to see them, feel they have not been at the scenes of violence especially when the assault was blunt force, then something is dreadfully wrong with their training. No program to the author's knowledge fails to include cast offs in the exercises, yet with some course graduates the training has not translated into recognition during actual investigations later. This does not mean the instructors were wrong, but it might mean they conveyed something that they did not intend to convey. A possibility is the overemphasis on subjective velocity impact spatter identifications, especially as classified on the basis of measured spatters alone. Size ranges are part of identification but must not be the only, and perhaps not even the most important, criteria.

The good news is that arterial damage patterns are recognized with a greater frequency now than they were 25 years ago. At that time in the past one rarely encountered case presentations involving arterial spurts. In recent years many cases in publications and shown at conferences identify arterial damage, although some still miss this important pattern identification.[1] In the past it was believed that only medical doctors identify whether or not the bloodstains at a crime scene are from arterial projection. In practice it is the experienced bloodstain pattern analyst who sees the arterial damage patterns first then contacts the pathologist to obtain verification. To date it is not required of autopsies to list arterial breach if it is not within the direct manner or cause of death. The guidelines were made before there were bloodstain pattern analysts who needed to have verification that arterial involvement existed. Hopefully guidelines and requirements in the future will make such notations mandatory.

Perhaps the greatest detriment to the future of BPE is a current attitude toward excluding the subject in the forensic science lab. This comes from crime lab directors, who themselves may not have had training in the subject. The belief is that the field is all police work and can be jettisoned in order to economize. What is happening, however, is that prosecutors are having to hire outside experts to carry the science testimony in supporting their law enforcement officers. This is not an economical approach. Furthermore not using the training in the laboratory with trace, firearms, clothing, and DNA is a waste of free and easily available information and corroboration.

SEMANTICS AGAIN

Traditional terms were developed within the context and knowledge of the time. Most of them have a basis in studies began in Europe and brought to the United States after the 1940s. There was nothing wrong with the terminology at the time. Unfortunately that was prior to 1940 and it is now 2007. The early terms

---

[1]Steel, Jennifer H, Lecea, Julie A., and Rocha, Elizabeth A., "The Effect of Speed on Bloodstain Patterns Found on the Exterior of a Moving Vehicle." Paper delivered at the 59th Meeting of the AAFS in San Antonio, Texas 2007.

were developed with input of scientists. Recent terms are being developed by law enforcement officers. There are problems with some science nomenclature in the changes.

An example is the use of the word passive instead of the old term low velocity impact spatter. Part of the confusion occurred when the impact site shifted from the target to the blood source. Paul Kirk labeled dripping blood as LVIS (low velocity impact spatter) because blood drops (Dr. Kirk used the word spatter for spots resulting from all the major dynamics) hit (impact) the target surface after being accelerated by gravity alone (a low velocity). It made sense in that context. When the position at which velocity was considered (e.g., impact site) shifted to the blood source from the target surface a problem developed. There was no impact to a blood source involved. When that was pointed out a number of alternatives came into being, gravitational, drip trails, low velocity stains, and passive stains. All of these are acceptable as long as investigators and those peripheral to the investigation know they all mean the same thing.

The term passive, however, is a misnomer and creates confusion regarding science labeling. The definition is that blood drops fall without any force acting upon them. Gravity is a force. Not only is gravity a force but it causes falling objects to accelerate. Using passive in a law enforcement context should be acceptable but it must be pointed out that in a science context blood drops falling by the force of gravity alone are not in a passive state. If a blood volume (pool) exists on a countertop, it is in a passive state. It has the potential to fall, rise (if hit), or flow (if the surface tilts). If it is just laying there, it is passive.

In a law enforcement context passive evidence may not be an active part of the crime. Bloodstains are never passive evidence. They are always an active part of the investigation, even if just to be eliminated as coming from violent or nonviolent acts. Bloodstains are the very definition of violent crime. It is incorrect to classify different bloodstain patterns as passive in order to make use of crime scene terminology, especially when it is in direct conflict with science terminology.

WHAT IS RIGHT RATHER THAN WHO IS RIGHT

Despite this argument regarding semantics, which terms are used in what context is far less important than answering the question of whether or not we are using the evidence to its fullest. There are many more applications that can possibly rival DNA and criminal profiling in the uniqueness of identifying people who commit violent crimes. Finding these new applications should be our focus with only side debates on terminology, instead of devoting the majority of time to "who is going to decide what we call it."

The author was initially trained by traditional methods and have retained some of the great information in that training. Meeting investigators around the globe has resulted in learning a variety of terms, most readily identifiable. Over a refreshing drink after the day is done, discussions regarding terms is invigorating exercise. It must not, however, be all we know and care about the evidence.

WHAT NEEDS CHANGING

In order to use BPA fully it is necessary to adequately understand all the permutations involved. Those who believe it is all bloodspatter analysis miss the full benefit of the evidence. Changing the title of it to bloodstain pattern evidence for the evidence and bloodstain pattern analysis for the expert review and testimony on behalf of the evidence has been encouraged since before 1983.

Certain principles, some misunderstood from basic training, need to be uniformly revised in literature and training formats:

1. There is no such thing as a standard drop size at crime scenes. For any given experiment the drop size can be standardized for the sample of blood used. This does not and cannot apply to actual crime scene evidence. The standardization of the size of a blood drop worked back in the 1970s to 1985 because out-of-date whole blood pouches were fairly uniform in hematocrit (red blood cell ratio). Maintaining this illusion now can create problems because it is difficult to find whole blood pouches. To deal with disease and because of hemorheology discoveries, packed red blood cells are universally used. Doing experiments with packed cells instead of whole blood will result in confusing variations on estimating a theoretical standard drop size. Using packed cells, however, creates bloodstains with the appearance of what is seen at actual crime scenes more than using whole blood. Perhaps this observation has been telling us something about the dynamics of crime.

2. Stop explaining blood behavior in terms of surface tension. Surface tension is a familiar feature. It explains why water drops hold together in the shape of drops. Water drops, however, wobble (oscillate) and form misshapen spheres in flight; blood drops do not wobble. Water has a much higher surface tension than blood, so obviously that isn't an explanation for the dichotomy in behavior. There is a reason for blood being a tight ball in flight, but water wobbles. Water is governed by surface tension, whereas blood is governed by cohesion. These are two different forces. We used to think surface tension explained all. This is not an embarrassment as a lot of extremely brilliant scientists once thought that all fluids behaved according to Newton's suggestions. It is now known that they don't so we change our view and move on.

    Some individuals may refer to the *Blood in Slow Motion* video[2] produced some years ago regarding blood drop oscillation. A close-up of a blood drop is shown with up and

---

[2]*Blood in Slow Motion* (1991). Video. Home Office Laboratory, London.

down movement and the voice-over describes the motion as oscillation. The drop has just separated from the blood source and the reaction to the break from the extension results in an up and down action/reaction. This occurs over less than 3 inches and does not include any of the wobble or sidewise motion associated with water under the same conditions. After the drop stabilized from separation it traveled without any distorted movement. The event is not an example of oscillation but rather an example of the stability and recovery of stability to separated blood drops.

3. Make use of medical research for input in expanding bloodstain pattern applications.   There is much information available from medical research. One does not have to be a medical doctor to have access to this. Funds are available for medical research, considerable funds in most countries of the world. Funds are not available for research in BPE. It makes economic sense to use the results of medical research to find information for the future. DNA came from funded programs in genetics, and toxicology information came from funded programs in pharmacology and chemicals. The list can go on. Bloodstain patterns are the primary focus of law enforcement, yet law enforcement in general does not fund studies in BPE.

4. Simply measuring spatters does not identify the dynamics that distributed them.   Measuring the spots that make up an arrangement falls far short of identifying the dynamics that created them. Individuals, who need tangible procedures to make identifications, shy away from understanding esoteric techniques. When attempts to memorize patterns prove unreliable they fall back on anything that can be explained with something factual such as mathematics. The individuals applying bloodstain patterns in this manner are not to be blamed, although what they claim justifies identification of HVIS and MVIS can cause extreme miscarriages in justice. The answer is to provide training in the use of specific criteria in identification, shape of the whole pattern, alignment of spatters, and distribution of spatters within the pattern (SAADD). This is a way to make what appears to be totally subjective into an objective approach. Other ways may be as good or better, but something needs to change in the training approaches to provide guidelines besides the size of spatters. There is too much size overlap, and noncriminal events can distribute the same size ranges.

5. Use team approaches for presentation of evidence.   A traffic accident investigator can take the witness stand and explain the dynamics of an accident without telling the trier of fact how a motor vehicle was designed and engineered. If vehicle engineering is necessary, the court may call an engineer to provide that information. The same should be true with bloodstain patterns. Law enforcement officers testify to what they know regarding the dynamics and distribution of the evidence in a crime context. If the jury needs to know how blood behaves in a scientific and reproducible manner the crime lab should provide this expertise. If injuries are a part of understanding the whole, the court can call a medical professional. Law enforcement officers are presently expected to provide information regarding the complete discipline, which should never have been laid on their shoulders. Much of the misconceptions encountered throughout the world has stemmed from attempts to carry out what has become required by the

courts. Conscientious division of responsibilities in investigations could resolve this for the benefit of all. Removing the technical explanations from the shoulders of some, but admittedly not all, officers could free them to concentrate on what they do best—record, collect, and investigate crime. Contributing to convictions, in some cases of innocent people, is not a validation of a technique as scientific.

THE MOST IMPORTANT ISSUE OF ALL—TRAINING

Efforts to standardize training has been on the agenda for decades. Unfortunately the standardization usually follows what has been traditional since before Paul Kirk's days. A new complete evaluation of training objectives and methods should be conducted. Consideration for segregating levels of training into need-to-know can be applied if all levels are available. Until that time 40-hour programs must meet the needs of a wide variety of academic levels and professional perspectives. This is difficult but not impossible if the correct objectives are in place.

RESEARCH IS ESSENTIAL TO ANY SCIENCE DISCIPLINE

The potential is there but not available until we have the benefit from true science research. The objective of research is to test theories and preconceived ideas in an open-minded way to see if they are in fact valid, and if they do apply as previously stated. Case examples do not satisfy research requirements if the objective is to prove a specific point. Workshops do not provide an environment to criticize what the instructor is teaching, and most instructors, author included, frown on efforts to do so.

Case reconstruction work can qualify as research if the results are analyzed without bias. The traditional method of reconstructing a crime scene by having all the same materials, carpet, wall paint, table and chairs, magazine issue, etc., will not be research quality if a bloody sponge is then hit and the results labeled as a pattern match based on both having spots. Distributing blood drops on the same surfaces does not constitute a reconstruction of the crime. On the other hand an exact reenactment of the crime may not produce the same recognizable patterns because of uncontrollable variables in the repeat.

SEARCH FOR COMMON GROUND BETWEEN DIFFERENT LEVELS OF TECHNICAL BACKGROUND

This is probably the whole conflict between those who disagree within organizations dealing with bloodstain pattern analysts. The evidence cannot be limited to nor specialized for one level of academic background. In the classes taught in

Sacramento we found all levels within each profession. Some law enforcement officers understood physics better than the PhD or MDs. Classes contained a delightful variety of viewpoints from PhD lab directors to college students who had never seen a crime scene. There is a need to recognize different levels of involvement with terminology as well as training.

Advantages of non-Newtonian behavior will eventually outweigh the inconvenience of the technical descriptions. Because the fluids are different from Newtonian, they reflect light differently and flow against the walls of capillary tubes differs. Stirring non-Newtonian fluids cause variations in behavior not found with Newtonian fluids. The way stains dry, clot, and mix with other fluids may differ. Most important of all, some blood vessels under certain circumstances will shift flow to Newtonian. When this happens the bloodstain patterns will differ in such a way that may make it possible to identify which injury projected which blood drop array. The information could be used to sequence assaults and corroborate and be corroborated by medical evidence. Hopefully this will all be much clearer in the future with BPE research.

The feeling of starting all over again is valid but need not be frustrating. DNA has revolutionized the way we look at crime investigations. It's unlikely that active investigators would vote to go back to the days of waiting for the sometimes equivocal results of serology, yet that was the way it was done at one time. Bloodstain pattern evidence belongs with DNA and other forensic science laboratory-based techniques. It will gain that level only if those who have access to actual crime scenes recognize the possibilities and start the chain of involvement by which other professionals come into contact with it.

APPENDIXES

FLOW DIAGRAM

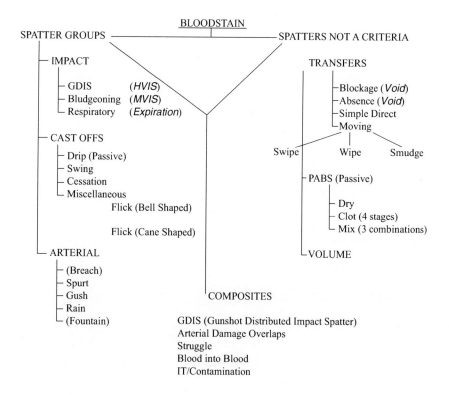

BLOODSTAIN

SPATTER GROUPS — SPATTERS NOT A CRITERIA

- IMPACT

  - GDIS   (*HVIS*)
  - Bludgeoning (*MVIS*)
  - Respiratory (*Expiration*)

- CAST OFFS

  - Drip (Passive)
  - Swing
  - Cessation
  - Miscellaneous

    Flick (Bell Shaped)

    Flick (Cane Shaped)

- ARTERIAL

  - (Breach)
  - Spurt
  - Gush
  - Rain
  - (Fountain)

TRANSFERS

- Blockage (*Void*)
- Absence (*Void*)
- Simple Direct
- Moving

Swipe   Wipe   Smudge

- PABS (Passive)

  - Dry
  - Clot (4 stages)
  - Mix (3 combinations)

- VOLUME

COMPOSITES

GDIS (Gunshot Distributed Impact Spatter)
Arterial Damage Overlaps
Struggle
Blood into Blood
IT/Contamination

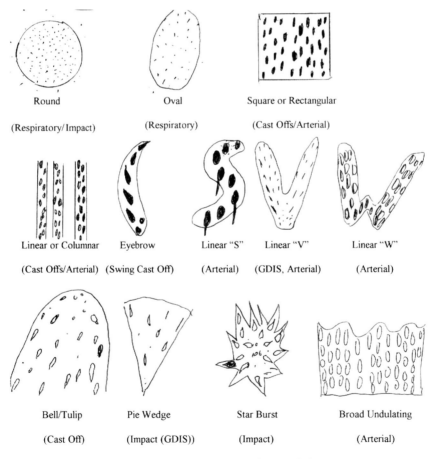

Round

(Respiratory/Impact)

Oval

(Respiratory)

Square or Rectangular

(Cast Offs/Arterial)

Linear or Columnar

(Cast Offs/Arterial)

Eyebrow

(Swing Cast Off)

Linear "S"

(Arterial)

Linear "V"

(GDIS, Arterial)

Linear "W"

(Arterial)

Bell/Tulip

(Cast Off)

Pie Wedge

(Impact (GDIS))

Star Burst

(Impact)

Broad Undulating

(Arterial)

Figure B-1   Shapes and probable distribution whole patterns.

AN OBJECTIVE APPROACH TO SPATTER CLASSIFICATION BASED ON SAADD

S signifies the shape of an arrangement of spatters (spots). This is the characteristic that takes the most experience and practice to become proficient in applying for classifications. The critical part of the observation is to recognize which spatters belong to only one rather than two or more different overlapping dynamic events. Be aware that overlapping patterns are expected at crime scenes. Some word descriptions regarding whole group shapes and the class with which they are most frequently associated are presented in Figure B-1.

Figure B-2
Directions of travel are toward the most irregular edge or narrowest point of tear drop shapes.

Directions of travel are the key to identification of a group of spatters as being part of a single dynamic act. A drop of blood leaving a spatter (spot) was traveling from the smoothest rounded edge in the direction of the most irregular, roughest edge of a bloodspatter. Because surface characteristics of the target may affect the length and width absorbency of blood fluid, an assumption that the longest measurement is always the direction of travel is unreliable. If all the stains within a group lack directions of travel, it may not be possible to conclude the dynamics that distributed them. So called "direct or right angled hits" of a group of blood drops resulting in spatters without directions of travel must be classified with the utmost care.

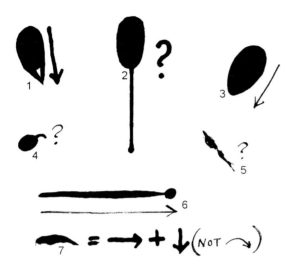

Figure B-3

Not all spatters (spots, bloodstains) may be interpreted as having directions of travel.

Figure B-4

Use of a piece of string to evaluate stain and pattern alignment.

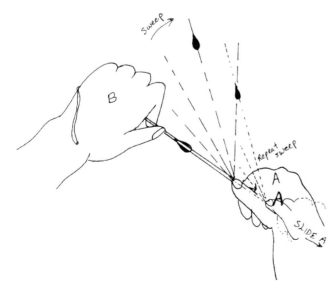

Stains 2, 4, and 5 in Figure B-3 do not have directions of travel. Stain 2 is a perfect oval. The tail seen is overflow from the quantity of blood deposited, not an indication of the direction the drop was traveling at impact. This occurs when multiple drops hit together or when the drop was exceptionally large (smaller drops did not separate). The contact marks for stain 7 are where the drop was recorded, while the tail is pulled toward gravity rather than as a direction of travel. Stain 4 is an enlargement of a fly regurgitated spot, not a direction of travel. Stain 5 resembles clot material, which is solid and unable to change shape to form a true spatter. The lumps of clot material may slip along a target surface leaving lighter shaded material behind, but this is unreliable in showing direction of travel.

AA is the abbreviation for alignment of spatters with respect to each other and with respect to the whole arrangement. Use of a piece of string helps identify spatters, which are at angles from each other and to locate a general position of the area of convergence.

Align the string in one hand (A) just prior to the base (first contact) of a stain. Extend the string in the other hand (B) along the length (axis of the stain) in the direction of travel. Sweep the string from side to side (with B) to see if it aligns with any other stains adjacent. If not, extend the length of the string back from the base (A), keeping it aligned through the middle of the original stain. Sweep again. Continue doing this until you have a minimum of five and preferably eight bloodspatters that appear to originate from a single common area. If you cannot find stains that are aligned with each other, consider the possibility that the pattern may not be an impact pattern. **Do not rely on bloodspatter sizes alone.**

Some examples of alignments and the classifications most frequently associated with them are presented in Figure B-5.

DD signifies density of spatters and distribution of size ranges. Impacts involve a variety of sizes from mist up to and including large stains. The important criteria is that all the sizes created from an impact are distributed from one location. This means that the drops travel outward at decreasing velocities since

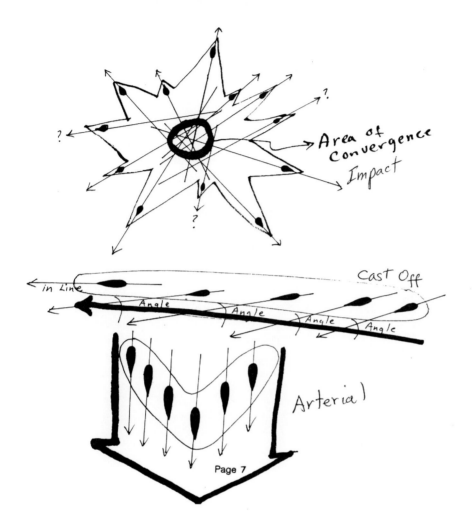

Figure B-5

Examples of alignment
of spatters with respect to
each other and the whole
pattern.

there is no input of additional momentum. Each drop will be deposited or fall by gravity when momentum is overcome or the drop meets a recording surface. Impacts are characterized by having many different sizes near the area of convergence (A of C) but diminishing to few different sizes as the pattern moves out away from the A of C.

Cast offs are sloughed off a moving object at different times and places over the range of movement. The object upon which the blood source is adhering and the force of centripetal motion will cast off stains in a relative size range and path of travel. There may be similar sizes or two or three different sized stains but they will be found distributed over the path of the object's travel. Like impact, however, cast offs usually do not have more blood added over the range. Patterns for cast offs tend to end more abruptly than impacts, but contain the same size ranges until they end.

Arterial damage projects blood from an injured blood vessel. The sizes may vary depending upon movement and the drop in blood pressure while the artery is open. The pattern, however, shows a remarkable uniformity of sizes over the whole arrangement. Unlike impact and cast off, arterial damage adds more blood over the range of the pattern. Thus the distribution and density remain relatively the same over the range of the pattern to the end. At the end of arterial damage patterns very large ovals overlapping may indicate marked drop in blood pressure, possibly the point in time of death.

Distribution for each of the three categories follow the dynamic acts. Impact spatters diminish with distance from the origin because all the spatters came from a common origin. When the array of drops, distributed at one moment in time, stop travel no more blood is available. Cast offs may also diminish in number but not as rapidly as impacts because new drops can be added from the object as it moves along the path. However, when the blood coating the object/carrier is gone, the pattern ends. With arterial distribution, drops are continually added, so density often remains the same over the entire range of the pattern and abruptly stops.

Some illustrations are presented in Figure B-6 of how density and distribution are interpreted for the three major classifications.

An important point to keep in mind is that exceptions to these do occur. Using several criteria helps to be confident that your analysis is accurate. Keeping these in mind when viewing the cases in Section II will help in learning to recognize what experienced investigators see in the evidence.

Figure B-6

Distribution of spatter sizes and quantity helps identify distribution events.

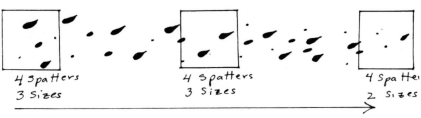

31 Spatters
5 Sizes

9 Spatters
3 Sizes

2 Spatter
1 Size

Distribution Suggestive of Impact Spatter Pattern

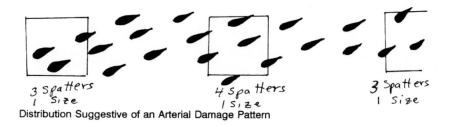

4 Spatters
3 Sizes

4 Spatters
3 Sizes

4 Spatter
2 Sizes

Distribution Suggestive of a Cast Off Pattern

3 Spatters
1 Size

4 Spatters
1 Size

3 Spatters
1 Size

Distribution Suggestive of an Arterial Damage Pattern

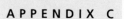

SPATTER IDENTIFICATION

The following table is a convenient grid for spatter identification. This can be reduced and carried as a wallet card.

Table C-1

Grid and Techniques for Impact Spatter Differentiation

Pattern	Force	Force Details	Flight Path Notes	Blood Source
GUNSHOT	Missile and expanding gases	Small contact area but high speed	Arc paths uncommon	Exposed at the moment of the event
BLUNT FORCE	Blunt object	Variable contact area, variable speed/medium energy common	Flight paths may include arcs	Must be exposed prior to impact
SPLASH	Blunt object	Large contact area with low energy	Always involves arcs on vertical adjacent surfaces	Must have pool accumulation prior to impact
BLOOD/BLOOD	Series of blood drops	Small to large contact area depending on time between drops and free-fall energy level	Always involves arcs on vertical adjacent surfaces	Requires rapid dripping source prior to impact
EXHALATION	Exhaled air	Compact oval or round contact area, variable energy level	May involve arcs but usually does not	Requires blood source in respiratory system

TWENTY-FIVE IMPACT BLOODSPATTERS

Measure and calculate the incident angles. Answers are on the next page.

ANSWERS

The guidelines for measuring bloodspatter is that the ratio is more important than absolute measurements. However, the width is usually easy to determine so that any variance will be the length. The following angles were determined with BPT (bloodstain pattern training) techniques because that method showed the best results with the blunt force impact mock crime scenes used in the Sacramento workshops. Completing the oval will provide results of higher values. The higher values are not a problem as long as the stains are on a horizontal surface. Problems may result if the complete-the-oval technique is used with the concept of points of convergence.

1. 30±3 (that is, 27–33 degrees are theoretically acceptable)
2. 26
3. 39
4. 63
5. 30
6. <10 (less than 10 degrees shouldn't measure)
7. 20
8. 34
9. 27
10. 34
11. 15
12. 27
13. 36
14. 66
15. 36
16. 18
17. 42
18. 41
19. 32
20. 23
21. 27
22. 70
23. <10
24. 22
25. 25

Stains 12 and 20 were from a cast off exhibit and are not impact spatters. All the others came from workshop exhibits from impacts only.

TIPS ON SEQUENCING PATTERN CATEGORIES

Beginning of the crime: patterns that may be involved in opening a blood source:

- Impacts
- Arterial beach and spurts

Middle of the crime: patterns involved in continued distribution of blood:

- Cast offs
- Arterial gushing, rain
- Transfers
- PABS/mix

End of the crime: patterns requiring a time period to form:

- Volume (pooling)
- PABS/clot, dry
- Flows

When a crime shows multiple areas of involvement, look for the smallest amount of blood and work forward. This works well when there are several investigators present because most of them will be at or near the end of the sequences where there is a lot more blood, and possibily a body. Be constantly aware of directions of travel. When dealing with multiple bodies, tracks (shoe, foot, drip cast offs) moving in two directions might indicate one assailant dealing with two victims. The lack of tracks between victims killed in different parts of the crime scene, as long as both victims appear to have been aware of the attack, could mean more than one assailant, i.e., one assailant dealing with each victim or one assailant keeping a group together while the other one murders. Exceptions to this have been encountered.

There is no such thing as always. General guidelines are beneficial for investigative leads, but exceptions occur. This is the main reason to not develop a scenario too soon. Be open to interpretations that contradict the first scenario.

Check deviations, don't ignore anything. Physical evidence may be more reliable than even multiple witness statements.

Sequence the crime scene before interviews. Once people have signed statements they are much less willing to admit they were wrong than if they know being less than honest will not be accepted, and may jeopardize their situation. It is essential, however, that the information gleaned from the scene be accurate. Confronting a suspect or witness with a scenario that they know is false provides an excuse to not be forthcoming with accurate details. In other words, if you start out wrong about what happened, witnesses and suspect may not correct you later.

Bloodstain pattern evidence can be essential in developing investigative leads. It is a loss of an investigative tool if the attitude persists to leave it all to the evidence technicians. Detectives who understand patterns can also benefit from the knowledge throughout the investigation.

SOME RANDOM VALUES FOR FUTURE RESEARCH[1]

HEMATOCRIT VALUES

Normal male 40–54%

Normal female 37–47

Children 35–49%

Babies, Newborn (via cord blood) 42–60%[2]

Values are available for babies through 6 months in Clinical Lab References.

FOR DIFFERENT ORGANS IN THE BODY FOR A PERSON WITH ARTERIAL HCT OF 43%[3]

Heart 20–25%

Brain 15–20%

Liver 40%

Lungs 35%

Spleen 80%

Kidneys 15–20%

VISCOSITY

Water (20° C) 1.0

Water (37° C) 0.7

Blood plasma (37° C) 1.2

Blood serum (37° C) 1.0–1.3

Whole blood (37° C) 2.0–10.0 (apparent viscosity)

Glycerol is 1000 times as viscous as water but is still Newtonian[4] in part because of inelasticity.

---

[1]Values may vary per geographical area. These were obtained online and from medical technology training notes in California.
[2]Tietz, Norbert W., Ed. (1983). *Clinical Guide to Laboratory Tests.* W.B. Saunders, Philadelphia, 258.
[3]Albert, Solomon N. Jain, Sumer Chand, Shibuya, Jo, Albert, Chalom A. (1965). *The Hematocrit in Clinical Practice.* Charles C Thomas, Springfield, 20–21.
[4]Barnes, H.A., Hutton, J.F., Walters, Kenneth. (1989). *An Introduction to Rheology.* Elsevier, Oxford, 2.

CLINICAL HEMOSTASIS VALUES

Capillary bleeding times 1–7 minutes (standardized blue lancet)

Clotting time <10 minutes

Clot retraction begins 1 hour and is maximum within 24 hours[5]

Traditional Lee White values of *in vitro* tube clotting is 5–8 min. for whole blood.

SEDIMENTATION RATE

0–20% within 1 hour

---

[5]Tietz, *Guide to Lab Tests,* 124.

GLOSSARY

The following definitions explain terms encountered in this text. Some are from terminology lists and others may be new and as yet unestablished in the lexicon. The latter are included for updating and simplicity of labeling. None of the definitions are written in stone.

Angle: See *Incident angle.*

Area of convergence: An area on a single plane surface from which at least three (preferably 5 to 8) bloodspatters have a common source of distribution. This location is in direct perpendicular line to the origin in space of the event from which the blood drops forming the bloodspatter arrangement were distributed.

Arterial damage stains: Bloodstains recorded by blood drops projected from a pressurized blood vessel.

Arterial fountain: An arterial damage bloodstain pattern recorded on a vertical surface and formed when blood is projected predominantly up, falling back by gravity, much like a fountain.

Arterial gush: An arterial damage pattern that forms when an extension column strikes a surface before it separates into drops.

Arterial rain: An arterial damage pattern that results from the fallout of an arterial fountain. These can be recognized as predominantly large stains scattered on a horizontal surface and lacking directions of travel.

Arterial spurt: An arterial damage pattern from a single extension of a column that has separated into drops before contacting a recording surface.

Back spatter: Usually used in the case of GDIS but actually applies to any impact. Blood drops separated from a blood source are distributed back toward the person holding the weapon.

Blood: A predominantly non-Newtonian, complex colloid, respiring physiological fluid found in vertebrates.

Blood into blood (B/B): When blood drips into a volume (pool) and satellite drops are distributed around the circumference. The amount and distance up and out that the satellite drops travel depend upon the frequency and force at which dripping occurs.

Bloodspatter: Some investigators use this term only to mean impact spatters or the stains that result when impact-distributed blood drops are recorded. Paul Kirk used the term for any spots distributed by events such as impact, cast off, and arterial projection.

Bloodstain: Evidence that fluid blood has come into contact with a recording surface.

Blunt force impact spatter (BFIS): A bloodstain pattern made up of blood drops distributed from a blunt force impact.

Bullet capsule blast: The small and fine blood drops (or resultant stains) found around a bullet, which were distributed with the bullet at exit from the blood source.

Caliper: Measuring device for bloodspatters.

Cast offs: Blood drops that are dislodged or sloughed off a moving object or carrier.

Clot: A physiological reaction that occurs in blood by which the liquid state changes to a solid stain with eventual extrusion of a changed liquid.

Compression pattern: See *Simple direct transfer.*

Contamination: Transfers and smudges to existing bloodstains at a crime scene by bystanders or emergency rescue teams.

Direction of travel (directionality): The direction a drop of blood was traveling when it became recorded on a surface. The direction is toward the most irregular edge, not always based on measurements.

Drip cast off: A cast off that was sloughed off a slow moving object, such as a person walking or running with an injury or carrying a bleeding victim or weapon. This is a suggested replacement for the traditional term *low velocity impact spatter.*

Entrance wound spatter (blow back, back spatter): Blood drop array distributed from the entrance wound of a projectile and the expanding gases that follow.

Exit wound spatter (forward spatter): The distribution of a blood drop array from the exit wound of a projectile.

Forward spatter: See *Exit wound spatter.*

Gunshot distributed impact spatter (GDIS): A pattern identified by two or more criteria, which may include almost point area of convergence, pie wedge overall, or cone arrangement of spatters; and small, fine, and mist included in the size ranges of bloodstains.

Hemolyze: A rupture of red blood cells when exposed to liquid of less specific gravity (saltiness) than the inside of the cells. Water is taken up by RBC to equalize conditions outside the cell. When more water is taken up than the cell can contain, the cell ruptures much like a balloon that has blown up too far. The result is a clear red colored liquid.

High velocity impact spatter (HVIS): A traditional term used to describe blood distributed by means of a high velocity force. The term was originally (previous to 1970) used to mean a drop of blood distributed at high velocity that left characteristic individual shaped bloodspatters.

Homologous diluent: A diluting fluid of the same or similar characteristics, such as saltiness, as that which is being diluted.

Impact site: A term that has become ambiguous in application. Originally it was used to mean the place where a blood drop was recorded with the velocity of the drop considered. This gave rise to the velocity impact spatter terms. After 1970 the term was shifted to mean the blood source (victim) where a forceful impact occurred. Confusion resulted when the old meaning was also retained.

Incident angle (angle of incident, angle of impact, impact angle): The acute angle between a blood drop flight path and the plane of a surface upon which it is recorded.

Inline beading: A distinctive beaded contact transfer pattern that may occur with natural hair. Oil along the hair shaft causes blood to bead. When the hair touches a surface the beads are recorded in a consecutive line of dots. Due to the absorbency of fabric, these can resemble spatters.

Inside angle: Within the acute angle from 0 to 90 degrees and contrasted with outside the angle, which is from 90 to 180 degrees. The acute angle is used at the tip of a bloodspatter for trigonometric functions. This has been equated to the incident angle on the basis of parallel lines intersected by a tangent (the target surface). Unfortunately no parallel lines exist in the context applied.

Investigative transfer: Patterns created by first line (or other) investigation personnel. This is contrasted with contamination, which is left by others assisting the victims. Investigative transfer should always be documented to avoid misinterpretations at a crime scene.

Low velocity impact spatters (LVIS): A term that is seldom encountered now but is another name for drip cast offs.

Medium velocity impact spatter (MVIS): Like HVIS, MVIS changed meaning after 1970. Presently it is applied when the crime is suspected or known to be by means of blunt force.

Moving contact stains: Transfer patterns that may have indications of directions of travel but have not been classified into swipe or wipe yet.

Muzzle blast: The rapidly expanding gases that hit a blood source opened by the bullet and cause atomization much like an artist spray brush assembly.

Origin: The position in space of an impact where blood drops separate from a blood source and are distributed away along flight paths.

Plasma: The clear or slightly opaque liquid portion of blood, which contains the factors involved in clotting (coagulation, hemostasis).

Parent drop: The larger bloodspatter that has acted as the source for a smaller drop projected ahead of it in the direction of travel.

Pattern: A form, shape, or outline made up of bloodstains from a single event or action. If comprised of spots (spatters) there must be adequate numbers to identify the whole pattern.

Physiologically altered blood stains (PABS): Bloodstains that retain characteristics that permit an analyst to recognize the physiological changes that occurred before the stain resulted.

Point of convergence/Point of origin: Originally terms of convenience referring to the drawing of directions of travel back from stains to a common conjunction. Some participants in workshops take the word "point" too literally and draw the stains until they converge at a point.

Pooling: Accumulation of a volume of blood over 1 cubic centimeter amounts.

Preblast: The small amount of gas that preceded the bullet and may atomize an already exposed blood source.

Protractor: A half-circle calibrated into degrees of angle.

Red blood cells (RBC): The discus-shaped cells that carry red pigment and give blood the red color. They carry oxygen from the lungs through the heart to the rest of the body and back.

Retraction (clot): The ability of blood to change from a liquid to a solid plus a liquid. The more advanced the retraction of the clot, the clearer and less pigmented the liquid around a solid clot.

Satellite spatter: Small- to fine-sized spatters found around the circumference of a larger stain.

Secondary spatter: Small drops that result from a parent drop hitting a rough target and distributing a second or subsequent drop in the direction of travel.

Serum: The clear liquid that remains after coagulation and red cell retraction. Upon exposure to light, serum will sometimes turn yellow, especially in cases of victims with liver damage from diseases such as alcoholism.

Serum stain: The shiny, shellac-like stains associated with redistribution of blood after clotting and retraction. Appreciating these may require oblique lighting and photographs.

Shadowing: The ability of blood to follow the path of least resistance (which is to avoid air) and curve around obstructions so that the edge of a blockage pattern looks like a shadow. This may also apply when an obstruction is removed before a spattering event ends, thus shading part of the blockage.

Simple direct transfer: Contact between a bloodied material and a nonbloodied recording surface.

Smear: See *Smudge.*

Smudge: A distortion of a moving contact stain that cannot be classified as to direction or act. This usually involves rubbing backward and forward.

Spatter: A bloodstain that has resulted from a blood drop. Some investigators use spatter as a shortened version of impact spatter to mean only the blood spots resulting from impact.

Spine: The pointed protrusions around the circumference of a drip cast off caused by rough surfaces breaking the cohesion of the drop on contact. These are marked directions of travel.

Splash: An event that happens when a low-level impact occurs with a volume (pool) of blood.

Splatter: Spatters distributed from a splash. Also a term used by individuals not adequately trained in bloodstain pattern analysis.

Spurt: See *Arterial spurt.*

String reconstruction: The crime scene technique of determining the origin of an impact by using angular estimates and strings to link the stains to a common origin. Alternative methods include lasers and computer programs.

Swipe: A moving transfer pattern of a bloodied material onto a nonbloodied recording surface. RBCs are deposited. The lift-off edge is usually lighter than the first contact edge.

Target: A recording surface for bloodstain patterns.

Template (stencil): An object that acts as an outline or central blockage and results in the details of the object recorded when painted over or placed in the path of a spattering event.

Transfer pattern: When blood is the medium by which the identity of an object is recorded. This may be by blockage during a spattering event, by simple direct contact, or by moving contact.

Volume (pool): When blood accumulates. The amount of blood considered a volume or pool varies between investigators but can be stated generally as enough to involve splash and stay wet long enough to contribute to IT and contamination.

Whole pattern: The complete or almost all of a group spots (spatters) that comprise a pattern.

Wipe: A moving contact pattern where RBC are removed. The lift-off edge is often darker than the beginning edge.

LIST OF ABBREVIATIONS

The use of abbreviations to economize time and space is becoming a way of life in most professions. The following list includes terms used in this work as well as those that may be encountered with research of literature applicable to bloodstain pattern evidence.

AAFS: American Academy of Forensic Sciences

AAPAPMO: Austra–Asian Pacific Association of Police Medical Officers

A of C: Area of convergence

A of O: Area of the origin

AD: Arterial damage

AS: Arterial spurt

B/B: Blood dripping into blood

BPA: Bloodstain pattern analysts (or analysis)

BPE: Bloodstain pattern evidence

BPT: Bloodstain pattern training (refers to techniques taught in Bloodstain Pattern workshops)

CAC: California Association of Criminalists

CFSS: Canadian Forensic Science Society

CHP: California Highway Patrol

CLS: Clinical laboratory scientist

CSF: Cerebral spinal fluid

CSI: Crime scene investigation (those who perform CSI have various titles such as ID tech, etc.)

D of T (DT): Direction of travel

DNA: Deoxyribonucleic acid

EDTA: Ethylene diamine tetra-acetic acid

ESR: Erythrocyte sedimentation rate

FSS: Forensic Science Society (United Kingdom)

FSS (HOME): Forensic Science Service (Home Office); Government Forensic Lab of the United Kingdom

GDIS: Gunshot distributed impact spatter

GSW: Gunshot wound

Hct: Hematocrit

Hgb: Hemoglobin

HVIS: High velocity impact spatter

IABPA: International Association of Bloodstain Pattern Analysts

IAFS: International Association of Forensic Sciences

IAI: International Association of Identification

ID Tech: Identification technician

LVIS: Low velocity impact spatters

MVIS: Medium velocity impact spatters

NSW: New South Wales (Australia)

PABS: Physiologically altered blood stains

PCV (PC): Packed cell volume (same as Hct)

PD: Police department (also, public defenders)

P of C: Point of convergence

P of O: Point of origin

RBC: Red blood cells (also, erythrocytes)

RNA: Ribonucleic acid

UCB: University of California at Berkeley

UCD: University of California at Davis

WBC: White blood cells (also, leucocytes)

INDEX

AAFS, see American Academy of Forensic
    Sciences
Abbot, Burton, 18
Abbreviations, listing, 375–376
ABC, see Appearance, behavior, and context
Academic projects, see Research
Accidental shooting, see Confession
Accomplice, see Confession
AIDS, see Human immunodeficiency virus
American Academy of Forensic Sciences
    (AAFS)
    Advanced Bloodstain Pattern workshop, 38
    funding of research, 342
Angle, see Incident angle
Appearance, behavior, and context (ABC),
        blood identification in photographs,
        57–58, 163
Area of convergence
    blunt force assault, 210
    definition, 38, 369
    determination, 38–39, 42–43
    traffic shooting case study, 90
Arterial breach, workshop exercises, 296–301
Arterial damage stain
    definition, 4, 369
    recognition trends, 346
    shapes, 354
    workshop exercises, 296–301
Arterial fountain
    definition, 369
    workshop exercises, 298–301
Arterial gush
    definition, 67, 369
    Lindsay case, 67
    workshop exercises, 297–301
Arterial rain
    decapitation by gunshot, 137
    definition, 137, 369
    workshop exercises, 298–301
Arterial spurt
    decapitation by gunshot, 138
    definition, 58, 369
    traffic shooting case study, 89
    workshop exercises, 297–301
Automatic weapons
    adjudication and resolution, 163

background, 161
    blood sources identification, 161
    bloodstain pattern identification, 161–163
    objective interpretation, 163
Autopsy, see Pathology

Back splatter, definition, 369
Bathtub homicide
    adjudication and resolution, 216
    background, 213
    blood flow with and without agitation,
        214–215
    blood sources identification, 213
    bloodstain pattern identification, 213–214
    objective interpretation, 215
B/B, see Blood into blood
Beating case
    adjudication and resolution, 107
    background, 101
    blood sources identification, 101–102
    bloodstain pattern identification, 102–105
    objective interpretation, 105–107
BFIS, see Blunt force impact spatter
Blockage pattern
    examples, 13
    head gunshot wound case, 81
    overview, 12
    simple direct transfer, 13
    workshop exercises, 303–308
Blood, definition, 369
Blood into blood (B/B)
    definition, 52, 369
    traffic shooting case study, 89
Bloodspatter
    automatic weapons, 162
    definition, 370
Bloodstain, definition, 370
Bloodstain pattern training (BPT), technique,
        37, 40–41
Bloody wall case, 227–231
Blunt force assault, hidden face case
    adjudication and resolution, 211
    background, 209
    blood sources identification, 209
    bloodstain pattern identification, 210
    objective interpretation, 210–211

Blunt force impact spatter (BFIS)
  accomplice confession case, 191
  case study, 168, 171
  definition, 168, 370
  workshop exercises, 287, 292
Blunt object homicide
  adjudication and resolution, 170
  background, 167
  blood sources identification, 167
  bloodstain pattern identification, 168–169
  confession by murderer, 248–249
  objective interpretation, 169–170
BPT, see Bloodstain pattern training
Bullet capsule blast, definition, 370

Caliper, definition, 370
Cast offs, see Cessation cast offs; Drip cast offs;
      Swing cast offs
Cessation cast offs
  beating case, 106
  definition, 7
  dynamics, 9
  staged stabbing case, 127
  workshop exercises, 293–296
Chicken House IV, 48
Chord, deposited sphere, 34–35
Clot
  clotting time, 176
  definition, 370
  hemostasis reference values, 368
Clothing
  collection and handling
    loose garments, 242–243
    suspects, 242
    victim, 241–242
  examination in crime lab, 243–246
  practical exams, 320–323
CODIS, see Combined DNA Indexing System
Combined DNA Indexing System (CODIS),
      253
Compression pattern, see Simple direct
      transfer
Confession, elucidation by bloodstain pattern
      evidence
  accidental shooting
    background, 220
    blood sources identification, 221
    bloodstain pattern identification, 221
    objective interpretation, 221–222
    resolution, 222
  accomplice persuasion
    adjudication and resolution, 193–194
    background, 189
    blood sources identification, 189–190
    bloodstain pattern identification, 191–193
    objective interpretation, 193
  adjudication and resolution, 187
  background, 183
  blood sources identification, 183–184
  bloodstain pattern identification, 184–185
  blunt object homicide, 248–249

example, 219
  objective interpretation, 185–187
Contamination
  definition, 119, 370
  self-defense staged shooting homicide, 119
Crime lab
  clothing analysis, 240–246
  DNA testing, see DNA testing
  interagency cooperation, 237–240,
      246–248

Daubert v. Merell Dow Pharmaceuticals, Inc.,
      271–273
Deborah number, non-Newtonian fluids, 27
Decapitation by gunshot
  adjudication and resolution, 139
  background, 135
  blood sources identification, 135
  bloodstain pattern identification, 136
  objective interpretation, 137–139
Defense attorney, see Legal system
Direction of travel
  definition, 72, 370
  Lindsey case, 72
Distribution patterns, bloodstains, 354
DNA testing
  bloodstain pattern evidence incorporation
    adjudication and resolution, 255
    background, 251
    bloodstain pattern identification,
        254–255
  DNA properties, 250
  evidence value, 108
  objectives, 251
  polymerase chain reaction, 253
  sample sources, 254
  short tandem repeats, 252–253
Drip cast offs
  beating case, 103
  definition, 7, 370
  self-defense staged shooting homicide, 115
  small caliber gunshot wound to head, 142
  stabbing case, 97
  staged stabbing case, 125, 131
  suicide staging in family, 176, 179
  witnessed homicide, 204
  workshop exercises, 293–295
Drive-by shooting staging
  adjudication and resolution, 201
  background, 197
  blood sources identification, 197–199
  bloodstain pattern identification,
      199–200
  objective interpretation, 185–187
  reconstruction, 200–201, 218
  traffic accident investigation, 225
Drying time
  blood flow, 58–59
  blood on glasses, 115, 117–118
Dying Lion of Nineveh, 4
Dying Lioness of Nineveh, 4

EDTA, see Ethylene-diamine tetra-acetic acid
Entrance wound spatter
   definition, 80, 370
   head gunshot wound, 80
Ethylene-diamine tetra-acetic acid (EDTA),
      preserved blood samples, 60, 281
Exit wound spatter
   definition, 10, 370
   examples, 10–11
Expert testimony
   abuse, 269
   attorney communication prospects, 275
   bloodstain pattern evidence benefits for
      defense attorneys, 269–270
   consistency, 75
   finding experts and qualifications, 270–271,
      274
   rationale for hiring, 268
Expiration, definitions, 12

Fingerprint
   beating case, 106
   bloodstain pattern evidence comparison, 76
Flow diagram, bloodstain patterns, 353
Footwear evidence, see Shoe print
Forensic crime lab, see Crime lab
Forward spatter, see Exit wound spatter
Frye v. United States, 271, 273

GDIS, see Gunshot distributed impact spatter
Grubb, Vicky, 263
GSW, see Gunshot wound
Gunshot distributed impact spatter (GDIS)
   angle of contact, 34–35
   automatic weapons, 160–161
   definition, 10, 370
   examples, 10–11
   Internal Affairs shooting investigation, 156–157
   mock crime scene, 39
   self-defense staged shooting homicide, 111–112
   shapes, 354
   small caliber gunshot wound to head, 142
   suicide staging in family, 176–177
   traffic shooting case study, 89, 91
   workshop exercises, 286, 288–289
Gunshot wound (GSW), *see specific cases*

Hair transfer, see Transfer pattern
Head pivot, modeling of gunshot effects, 116,
     118–119
Hematocrit, see Red blood cell
Hemolyze, definition, 370–371
Hemostasiss, see Clot
High velocity impact spatter (HVIS), see also
     Gunshot distributed impact spatter
   automatic weapons, 160–162
   definition, 50, 371
   identification, 52
   small caliber gunshot wound to head, 142
   traffic shooting case study, 90, 92
   workshop exercises, 286, 288–289

HIV, see Human immunodeficiency virus
Homologous diluent, definition, 371
Hot Lips device, 334–335
How to Kill a Horse, 2
Human immunodeficiency virus (HIV),
     biohazard considerations in workshops,
     278–281
HVIS, see High velocity impact spatter

IABPA, see International Association of
     Bloodstain Pattern Analysts
Impact site, definition, 371
Impact spatter
   definition, 8
   velocity impact spatter terms, 8–10
Imwinkelried, Edward, 271–274
Incident angle
   calculation, 363–364
   definition, 371
Informant execution
   adjudication and resolution, 143
   background, 141
   blood sources identification, 141
   bloodstain pattern identification, 142
   objective interpretation, 142–143
Inline beading
   beating case, 103
   definition, 371
Inside angle, definition, 371
Internal Affairs shooting investigation
   adjudication and resolution, 157
   applications of bloodstain patterns,
     226
   background, 155
   blood sources identification, 155
   bloodstain pattern identification, 156
   objective interpretation, 156–157
International Association of Bloodstain
     Pattern Analysts (IABPA)
   case sharing, 329
   terminology standardization, 6
Investigative transfer
   definition, 371

Jury, see Legal system

Kirk, Paul Leland
   Abbot Case, 18
   biography, 17–18
   blood velocity versus volume observations,
     8–9, 34, 50–51
   impact site definition, 49
   photograph, 16
   research interest, 20
   Sheppard case, 8, 19
   terminology, 347

Law enforcement
   forensic science communication, 6
   police shooting, see Internal Affairs
     shooting investigation

Legal system, *see also* Expert testimony
  Australia bloodstain patterning evidence,
      226–227
  bloodstain pattern evidence benefits for
      defense attorneys, 269–270
  case material preparation for trial,
      56–59
  case resolution and justice, 206
  Daubert v. Merell Dow Pharmaceuticals, Inc.,
      271–273
  juror attitudes, 272
  physician intimidation, 259
  scientist's opinion of law
      dichotomy of viewpoints, 265–266
      differences between law and science,
          266–268
Lindsey, Alexander
  adjudication and resolution of case, 73,
      226–227
  blood sources identification
      bleeding injuries, 66
      prior wet bloodstains, 66–67
  bloodstain pattern identification, 67–70
  case background, 64–65
  motive examination, 74
  objective interpretation, 70–73
  second inquiry origins, 63–64
Low velocity impact spatter (LVIS)
  definition, 7–8, 371
  head gunshot wound, 80
LVIS, see Low velocity impact spatter

Medium velocity impact spatter (MVIS), *see
    also* Blunt force impact spatter
  beating case, 105
  bloody wall case, 231
  causes, 75
  definition, 50, 371
  identification, 52
  workshop exercises, 287–288
Mist
  spatter and resolution, 10–11
  workshop exercises, 287–289
Moving contact stain, definition, 371
Muzzle blast, definition, 372
MVIS, see Medium velocity impact spatter

National Institute of Forensic Science (NIFS),
    227
Newton, Isaac, 23–24
Newtonian fluids
  axial flow, 28–29
  behavior, 23–26
  blood drops, 25
  crime event influences, 29–30
  frictional stress per unit area against vessel
      wall, 26
  meniscus, 31
  reflected image, 31
NIFS, see National Institute of Forensic
    Science

Non-Newtonian fluids
  axial flow, 28–29
  blood
      behavior, 24–26
      composition effects, 29
      physiological significance, 29–30
  crime event influences, 29–30
  definition, 24
  flow, 28
  frictional stress per unit area against vessel
      wall, 26–27
  meniscus, 31
  prospects for study, 351
  reflected image, 31

Origin
  definition, 372
  reconstruction, 44–46

PABS, see Physiologically altered blood stain
Parent drop, definition, 372
Pathology, bloodstain pattern evidence
      incorporation
  advantages, 257–258
  autopsy report errors, 258–259
  case study
      adjudication and resolution, 262
      autopsy findings, 260–261
      background, 260
      blood sources identification, 260
      bloodstain pattern identification, 260
      objective interpretation, 261
  external evidence notation, 258
Pattern, definition, 372
PCR, see Polymerase chain reaction
Penal system, bloodstain patterning evidence
      utilization, 231–232
Photography
  blood identification, 57–58
  evidence technician tips, 232–234
  work-ups for training, 316
Physiologically altered blood stain (PABS)
  bathtub homicide, 213
  definition, 372
  shoe examination, 238–239
  workshop exercises, 308–312
Physiologically altered bloodstains
      (PAB)/clot
  beating case, 102
  definition, 81
  staged stabbing case, 127
  suicide staging in family, 176
Physiologically altered bloodstains (PAB)/
      mixed-cerebral spinal fluid
  assault positioning reconstruction,
      192–194
  definition, 66–67
  source, 257–258
Plasma, definition, 372
Plea bargain, see Witnessed homicide
Point of convergence, definition, 372

Point of origin
   definition, 372
   reconstruction, 44–46
Police, see Law enforcement
Police shooting, see Internal Affairs shooting
      investigation
Polymerase chain reaction (PCR), DNA
      testing, 252–253, 253
Pooling
   definition, 372
   pattern, see Volume
Preblast, definition, 372
Prison, see Penal system
Prosecutor, see Legal system
Protractor, definition, 372

Red blood cell
   blood content in non-Newtonian fluids,
      29–31
   hematocrit
      organ differences, 367
      reference values, 367
      variations, 35
Repositioning, bodies after staged suicide,
      178–180
Research, bloodstain pattern evidence
   academic projects
      conducting study, 343
      data, variable, and control identification,
         333–334
      device design and manufacture, 334–337,
         340–342
      evaluation and conclusions, 343
      funding, 342
      initiation steps, 330
      literature search, 332–333
      secure area for study, 334
      subject selection, 330–332
   importance and prospects, 350
   overview, 327–329
   reconstructions, 328–329
   workshop value, 327–328
Resolution, bloodspatter, 11
Respiratory patterns
   beating case blood pattern, 103
   expiration definitions, 12
   Hot Lips device, 334–335
   shapes, 354
Retraction, definition, 372
Reynold's number
   Deborah number, 27
   frictional stress per unit area against vessel
      wall, 26

SAAD, see Shape, alignment, arrangement,
      density, and distribution
Satellite spatter
   definition, 372
   head gunshot wound case, 81
   Newtonian fluids, 30
   self-defense staged shooting homicide, 113

Secondary spatter, definition, 372
Sedimentaton rate, blood, 368
Self-defense staged shooting homicide
   adjudication and resolution, 117–119
   background, 111
   blood sources identification, 111
   bloodstain pattern identification, 111–114
   objective interpretation, 114–117
Sequencing
   events at crime scene, 365–366
   traffic accident investigation, 223–225
Serum
   definition, 372
   stain, 372
Shadowing, definition, 373
Shape, alignment, arrangement, density,
      and distribution (SAAD), spatter
      classification, 355–359
Sheppard, Samuel, 8, 19
Shoe print
   dry blood spatter print, 84
   practical exams, 321–322
   self-defense staged shooting homicide, 114,
      116
   shoe identification, 238–239
   staged stabbing case, 130
Shooter identification case
   adjudication and resolution, 84
   background, 79
   blood sources identification, 79–80
   bloodstain pattern identification, 80–81
   objective interpretation, 81–84
Short tandem repeats (STRs), DNA testing,
      252–253
Simple direct transfer, definition, 13, 373
Slow Bullet device, 290, 335
Smear, see Smudge
Smudge
   definition, 305
   workshop exercises, 305–308
Sneeze, beating case blood pattern, 103
Spatter
   definition, 373
   grid for identification, 361–362
   SAAD classification, 355–359
Spine, definition, 373
Splash
   definition, 373
   self-defense staged shooting homicide, 111,
      114
Splatter, definition, 373
Spurt, see Arterial spurt
Stabbing cases
   adjudication and resolution, 98
   background, 95–96
   blood sources identification, 96
   bloodstain pattern identification, 96–97
   murder staging in family
      adjudication and resolution, 247–248
      background, 246
      blood sources identification, 246–247

Stabbing cases (*Continued*)
  bloodstain pattern identification and
    serological analysis, 247
  objective interpretation, 97–98
  staged case
    adjudication and resolution, 131
    background, 123–124
    blood sources identification, 124
    bloodstain pattern identification,
      124–128
    objective interpretation, 128–131
Staged crimes, see Bathtub homicide; Drive-by
    shooting staging; Self-defense staged
    shooting homicide; Stabbing cases;
    Suicide staging in family
Standardization
  goals, 5
  terminology, 6, 347
String reconstruction, definition, 373
STRs, see Short tandem repeats
Suicide staging in family
  adjudication and resolution, 180
  background, 173
  blood sources identification, 173–174
  bloodstain pattern identification
    first victim, 174–175, 178–179
    second victim, 175–177, 179–180
  objective interpretation, 177–180
Swing cast offs
  accomplice confession case, 190, 192
  definition, 4, 76, 370
  dynamics, 8
  shapes, 354
  stabbing case, 97
  workshop exercises, 293–296
Swipe
  blunt force assault, 210
  definition, 373
  head gunshot wound case, 81
  stabbing case, 97–98
  workshop exercises, 305–308

TAI, see Traffic accident investigation
Target, definition, 373
Template, definition, 373
Terminal velocity, blood drops, 35
Traffic accident investigation (TAI)
  adjudication and resolution of case, 59–60
  blood sources identification, 53–54
  bloodstain patterns
    identification, 54–55
    consistency with traffic accident, 59
  case material preparation for trial, 56–59
  case overview, 52–53
  objective interpretation, 55–56
  sequence of impact determination, 223–225
  staged homicide case study, 52–61
Traffic shooting case study
  adjudication and resolution, 92
  background, 87
  blood sources identification, 87–88

  bloodstain pattern identification, 88–90
  objective interpretation, 90–91
Training, bloodstain pattern evidence
  Advanced Bloodstain Pattern workshop, 38
  arterial damage exercises, 296–301
  blood handling, 281–282
  blood products, substitutes, and biohazard
    considerations, 278–281
  bloodstain pattern training technique, 37,
    40–41
  cast off evidence exercises, 292–296
  clean-up techniques, 282–294
  demonstrations, 323
  exams
    final exams, 324–325
    overview, 315
    practical exams, 320–323
  impact spatter exercises, 287–292
  importance and prospects, 350
  pattern puzzles, 316
  photo work-ups, 316
  physiologically altered bloodstains, 308–312
  recommendations, 41–44
  reconstruction exercise, 316–318
  research value, 327–328
  spatter practice, 318–320
  student observations, 277–278
  transfer pattern exercises, 303–308
  volume bloodstain patterns, 3122–313
Transfer pattern
  definition, 373
  hair transfer
    beating case, 105
    blunt force assault, 211
  staged stabbing case, 126, 129
  workshop exercises, 303–308
Trial, see Legal system
Trigonometry
  assumptions and problems in blood pattern
    evidence studies, 33–36
  chord of deposited sphere, 34–35
  measurement techniques and blood drop
    velocity effect studies
    accuracy concerns, 42–43
    Advanced Bloodstain Pattern workshop,
      38
    areas of convergence, 38–39, 42–43
    bloodstain pattern training technique,
      37, 40–41
    California Highway Patrol, 36–37
    recommendations for training, 41–44
  origin reconstruction, 44–46
  sphere projection, 36

Urination, beating case, 102–104

Velocity impact spatter (VIS)
  terminology, 50
  variables for identification, 50–52
VIS, see Velocity impact spatter
Viscosity, reference values, 367

Void, definition, 12
Volume
    definition, 24, 373
    small caliber gunshot wound to head,
        141–142
    staged stabbing case, 130
    suicide staging in family, 178
    workshop exercises, 312–313

Water–blood interactions, see Bathtub
        homicide
Whole pattern, definition, 374
Wipe
    definition, 374
    stabbing case, 97–98
    workshop exercises, 305–308

Witnessed homicide
    adjudication and resolution, 151
    background, 147–148
    blood sources identification, 148
    bloodstain pattern identification,
        148–150
    objective interpretation, 150–151
    plea bargain case in homicide
        adjudication and resolution, 205–206
        background, 203
        blood sources identification, 203–204
        bloodstain pattern identification,
            204
        objective interpretation, 204–205
    witness unreliability, 152, 203–206
Workshops, see Training